WHAT INSECTS DO, AND WHY

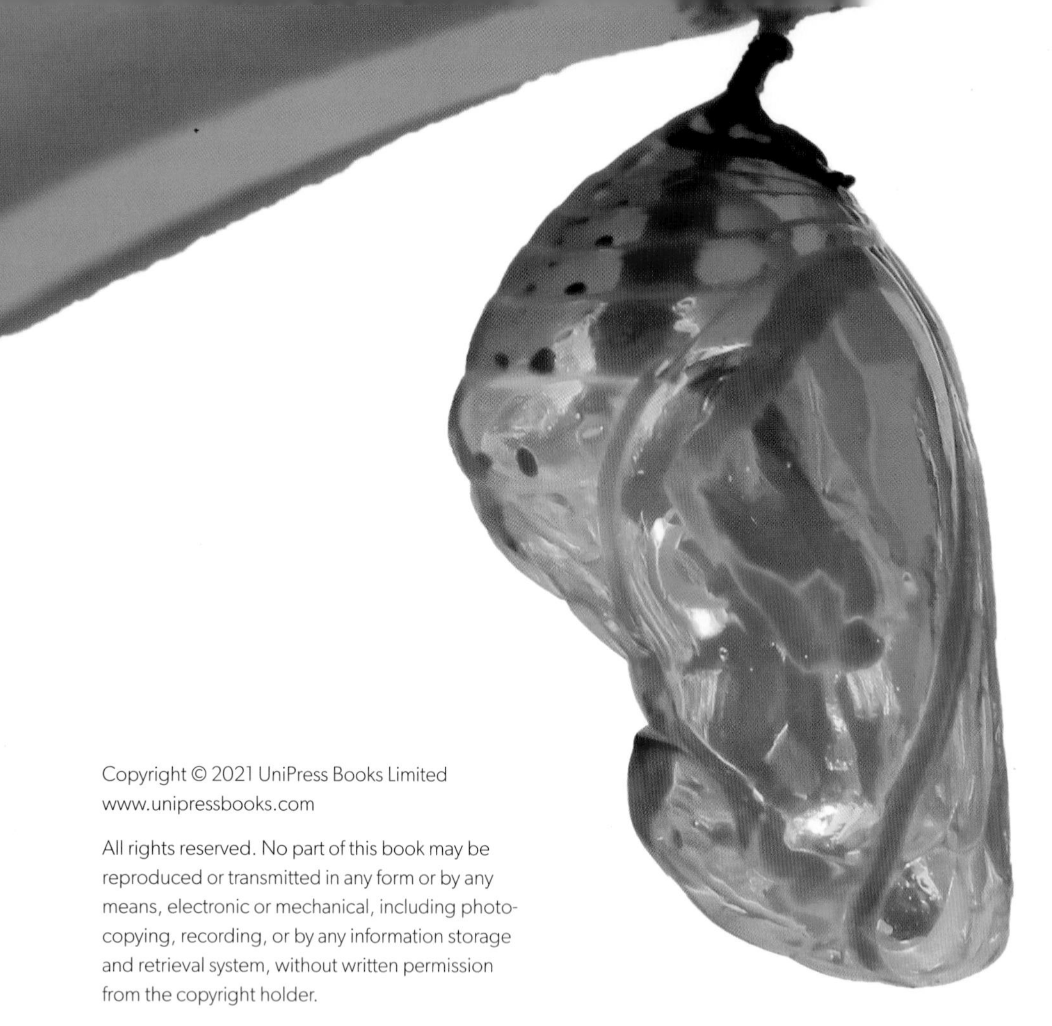

Published in the United States and Canada
in 2021 by
Princeton University Press
41 William Street, Princeton, New Jersey 08540
press.princeton.edu

Publisher: Nigel Browning
Commissioning editor: Kate Shanahan
Project manager: Natalia Price-Cabrera
Designer & art director: Paul Palmer-Edwards
Picture researcher: Natalia Price-Cabrera
Illustrator: Sarah Skeate

Library of Congress Control Number 2020952248

ISBN 978-0-691-21769-7

Printed in China

10 9 8 7 6 5 4 3 2 1

WHAT INSECTS DO, AND WHY

ROSS PIPER

PRINCETON UNIVERSITY PRESS
PRINCETON AND OXFORD

CONTENTS

INTRODUCTION 6

1 LIFE CYCLES 30

2 TO EAT AND BE EATEN 58

3 DEFENSES 94

4 SOCIALITY 120

5 PARASITOIDS AND PARASITES 156

6 QUID PRO QUO 182

INSECTS IN A CHANGING WORLD . . . 206

FAQs 212

GLOSSARY 218

INDEX 220

ACKNOWLEDGMENTS 223

INTRODUCTION

I MIGHT BE BIASED, but insects are the most fascinating animals with which we share this planet. From my first tottering steps I was spellbound, which was bad news for the various small beasts in my yard and wider neighborhood. Small, grubby fingers ferreting unfortunate beetles and caterpillars from their hiding places and incarcerating them in plastic tubs, often together and often in pretty squalid conditions. In my defense, it was the early 1980s, I was young and I didn't know any better. My favorite was a violet ground beetle. At the time, I had no real idea what it was. All I knew was that it was big, metallic purple, and that it deserved a stint in a tub. This is probably how it begins for all insect-botherers.

When you take time to find and observe these animals, you begin to understand their fabulous diversity. Their appearance is so varied and often so strange that they make sci-fi monsters look a bit lame. Then there are the ways in which they live. From microscopic, incestuous wasps that live their whole life in the eggs of other insects, caterpillars that fool ants into believing they are their sisters, and beetles that lure flies to their doom using stinky secretions. If you want sex, violence, and intrigue, insects have all of this and much more besides.

Their lives are so juicy that you could easily fill ten meaty tomes with what insects get up to, yet we have only scratched the surface of understanding how they live. To date, just over one million species of insect have been described, but there are still millions more out there awaiting description. Our knowledge of even the described species is generally very poor. For the vast majority of insect species, we know next to nothing about the ins and outs of their lives. Insects tend to get overlooked because of their generally small size. As well as being overlooked, they are generally maligned animals because a few species nibble our crops, run amok in our homes, or transmit diseases to us, our pets, and livestock.

We are all familiar with insects, but what are they? Let's begin at the beginning. Insects are animals. You, a scuttling beetle, and a sea anemone all share a common ancestor that lived probably one billion years ago. I've lost count of the times I've read or heard "animals and insects." If you see or hear this then it is your duty to correct it. Within the animals, insects are a type of arthropod.

What Is an Arthropod?

Arthropods are the animals we commonly call "bugs" or in even more derisory terms, "creepy crawlies." This is the largest group of animals by far (1.25 million known species and counting), way bigger than all the other animal groups

∨ This phylogenetic tree shows how the different groups of arthropods are related. This reveals that insects are crustaceans.

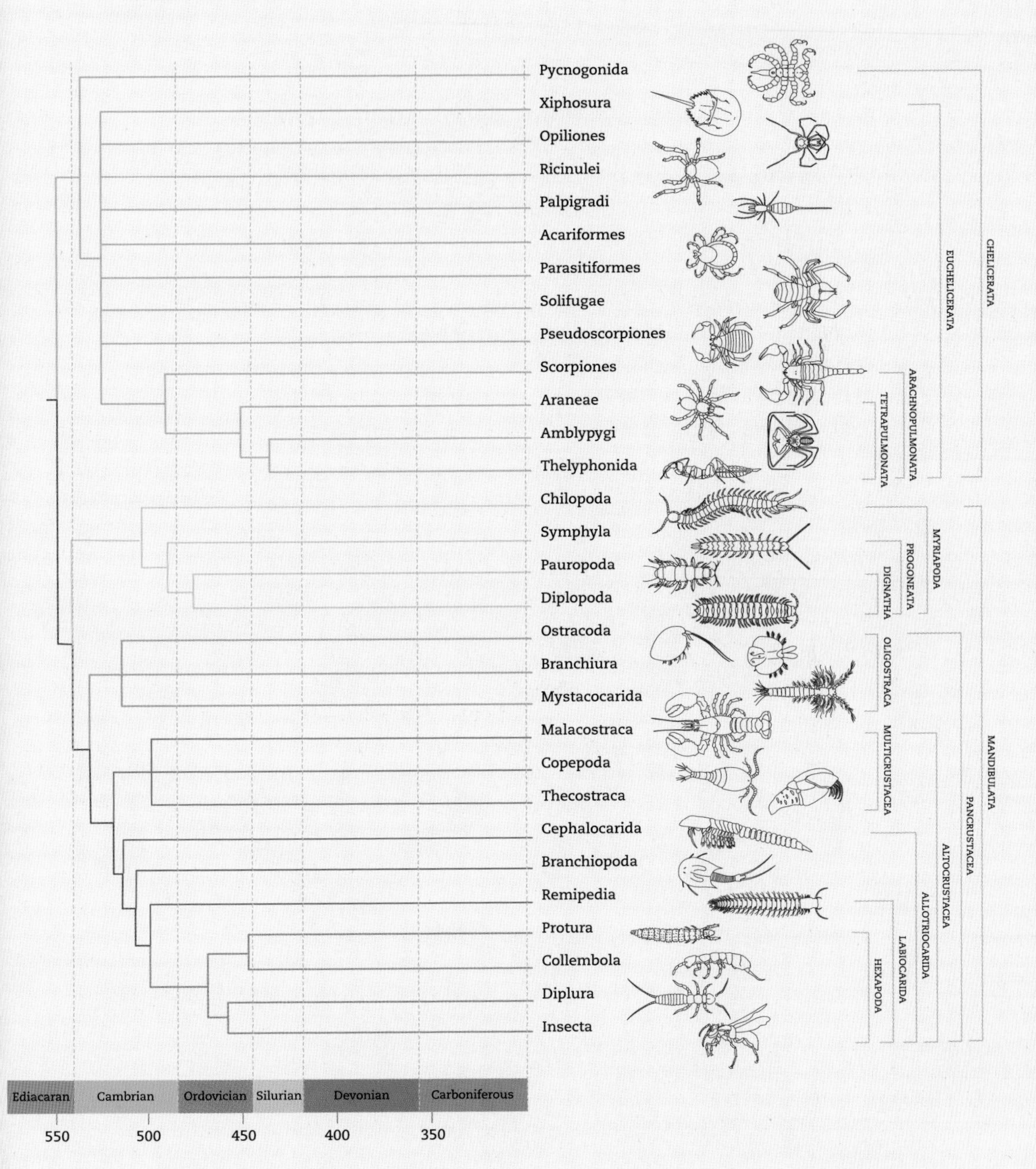

combined. The arthropods are the insects, springtails, and their relatives, crustaceans, arachnids, horseshoe crabs, sea spiders, centipedes, millipedes, and their relatives. This is a mind-bogglingly diverse group of animals, but they all share the following features:

- Jointed limbs
- An exoskeleton made of chitin, often reinforced with calcium carbonate
- The exoskeleton must be periodically shed for the animal to grow
- A segmented body; each segment often has a pair of appendages.

From an evolutionary point of view, the insects are terrestrial crustaceans. Their closest living relatives are two enigmatic groups of crustaceans—the remipedes and the cephalocarids. The former are rarely seen denizens of flooded caves and the latter are tiny animals that inhabit marine sediments. Insects also go back a very, very long way, with their likely origins some 480 million years ago in the early Ordovician Period. Between then and now planet Earth has been many different worlds, but the insects took to a life on land and made it their own.

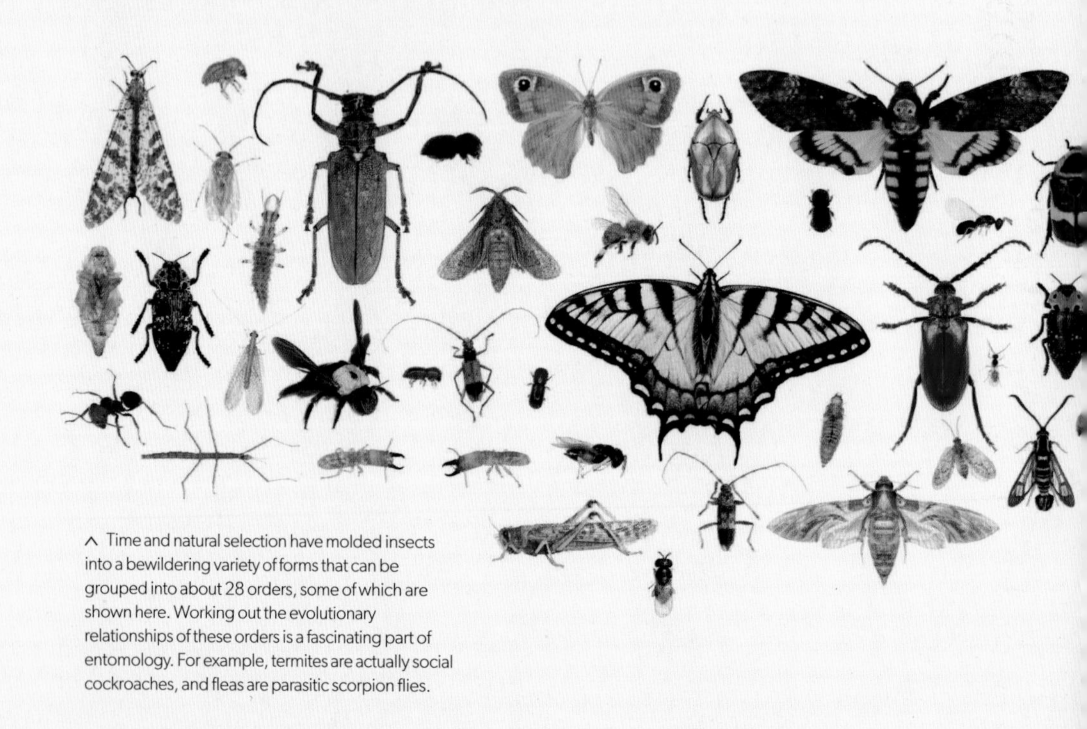

^ Time and natural selection have molded insects into a bewildering variety of forms that can be grouped into about 28 orders, some of which are shown here. Working out the evolutionary relationships of these orders is a fascinating part of entomology. For example, termites are actually social cockroaches, and fleas are parasitic scorpion flies.

What Is an Insect?

In insects, the general arthropod body-plan has been tweaked and they all share the following features:

- A three-part body: head, thorax, and abdomen
- Compound eyes and often simple eyes, too
- One pair of antennae
- Wings (secondarily lost in some insects).

During the long history of these animals, this form has been fine tuned by time and the environment into the extraordinary diversity of forms we see in living insects today.

By any measure, insects are among the most successful land animals there have ever been. They exist in such numbers and live in such bewildering ways that they pull at every single thread of the terrestrial domain. What are the secrets to this success?

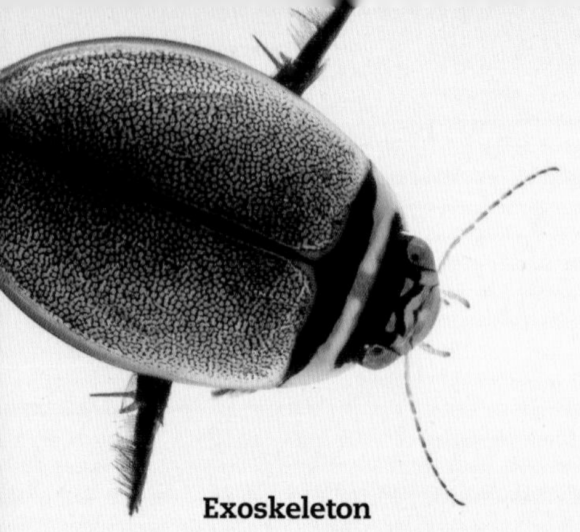

< Aquatic insects have a smooth exoskeleton and paddle-like limbs.

> Top: Adapting to life among running water has led to the evolution of flattened, hydrodynamic forms, such as this mayfly nymph shown here.

Bottom: *Halobates* water striders live on the surface of the open ocean. Their exoskeleton is exceedingly water repellent.

Exoskeleton

First, is the exoskeleton. This is so much more than a suit of strong, light armor. It is the insect's skeleton, composed of chitin, and so it must provide anchor points for all the animal's muscles, which it does via tiny, inward projections. It must keep the insides in and the outside out. Crucially, it needs to prevent undue water loss as dehydration is the ultimate challenge for all terrestrial beings. A waxy layer prevents the insect from becoming a dried-out husk. Insect exoskeletons range from the incredibly delicate, such as that sported by ephemeral beings like mayflies, to the almost impregnable, such as those of the aptly named ironclad beetles and some weevils. The exoskeleton and the efficient musculature beneath made life on land possible for the first insects, but it also has one major flaw. It's not very elastic. So, when an insect needs to grow it has to escape its old exoskeleton and make a new one. This sounds rather straightforward, a bit like a snake shedding its skin, but it is so much more complicated. For starters, the

∨ The earwig on the left has just shed its exoskeleton. The new exoskeleton is pale and soft.

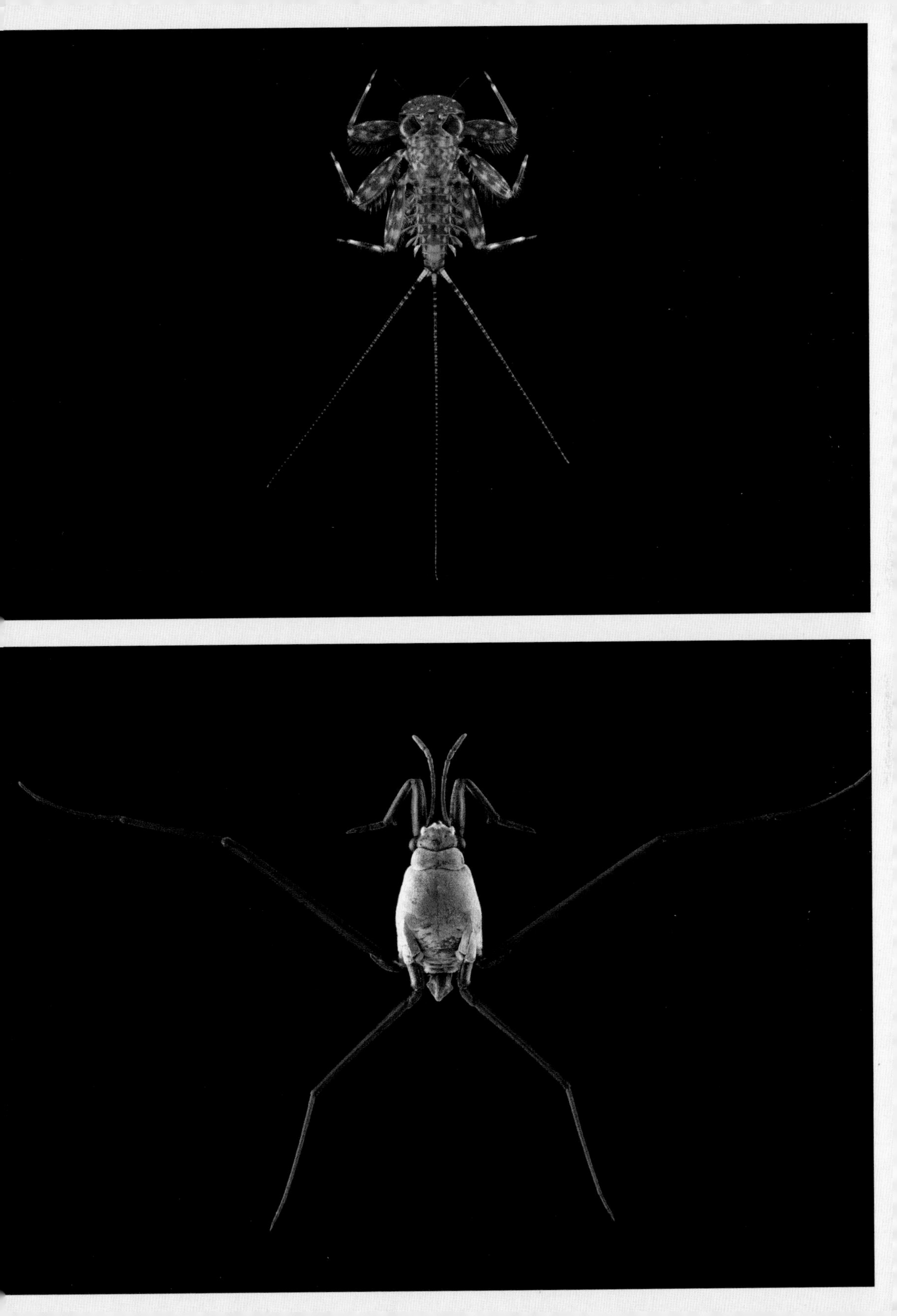

< The limbs of insects come in a bewildering array of forms, such as the rear legs of *Calodromus mellyi*, which are used in defense and perhaps courtship.

^ The exoskeleton is most armor-like in the beetles, such as this *Leptochirus* rove beetle.

exoskeleton extends down the insect's throat to line its foregut and up through the anus to line the hindgut. The exoskeleton also lines the vanishingly tiny tubes that transport gases to and from all the insect's tissues. All this has to be replaced too for the insect to grow. The whole process is a remarkable natural phenomenon precisely controlled by a symphony of hormones. The new exoskeleton has to be synthesized and ready before the old one is cast off, and the physical process of leaving the old exoskeleton is fraught with danger. Not only that, but when the insect is finally free of its old covering, the new one is still soft and pliable, leaving the animal acutely vulnerable. Considering all of this, it's a wonder that any animal with an outer layer like this could ever be more than just a footnote in the history of life on Earth, but its downsides are more than made up for by the protection and opportunities it provides an insect.

If you have a good look at a range of insects you'll be struck by their colors. They come in a staggering variety of hues, from the vivid red of a lily beetle to the deep, metallic iridescence of a long-legged fly. These colors can be formed by pigments, but in those insects with metallic iridescence the colors are due to the scattering of light by the crystalline structure of the exoskeleton. To our eyes, lots of insects look like living jewels.

The exoskeleton is also festooned with all manner of outgrowths that appear as hairs, bristles, and scales. These structures are multifunctional. The wings of butterflies are covered in huge numbers of scales that give the wings their colors and patterns. These scales can be rather fur-like providing their owners with insulation and protection from predators. Many beetles are dusted with scales of every hue that are easily dislodged.

The exoskeleton also sheathes the jointed appendages of the insect, which come in every shape and size. The appendages of the head have been modified into all manner of shears, syringes, and saws for making short work of food. Alongside these mouthparts are delicate little limbs called palps that mainly taste and manipulate food. Emanating from the head are a single pair of antennae. In some insects these are almost invisible, but in others they're fantastically elaborate and bristling with sensory pits for detecting food and members of the opposite sex. The limbs of insects have the same basic structure, the segments of which are named after the structures of vertebrate legs, so they have a coxa, trochanter,

∨ Outgrowths on the exoskeleton include all manner of bristles and scales, such as the fur-like covering of many moths (*Deilephila elpenor*).

< The mandibles of this nut weevil are at the tip of its long snout-like rostrum (*Curculio nucum*).

femur, tibia, and tarsus. The tarsus is normally tipped with a pair of claws. Natural selection has worked wonders on this basic limb form. For example, you can see lots of unrelated insects equipped with raptorial legs for snatching prey.

In others, such as the mole crickets, the legs are beautifully adapted for digging, while others have beefy rear legs that power prodigious jumping abilities. Some leaf hoppers are even equipped with cogs on the upper parts of their legs that keep the legs perfectly in time when they jump. If you've ever seen a flea beetle or one of these leaf hoppers in action you'll understand that insects are the undisputed champions of jumping.

Wings and Flight

The most significant extension of the insect exoskeleton and the innovation that makes these animals so remarkable is the wing. Have a good look at an insect wing—under a microscope if you can. They're one of the most elegant structures in nature. Watch a hoverfly in the summer months and marvel at what they do with these wings—they make other flying animals look a bit clumsy. The level of precision that goes into their flying far outstrips that of most larger flying animals.

Wings appeared very early in the evolution of the insects, probably around 400 million years ago, which is at least 170 million years before vertebrates ever took to the air. The origins of the insect wing are hotly debated, although recent research suggests they evolved from legs. Regardless of their origin, this innovation completely transformed the fortunes of the insects. As the wings and their musculature became ever more fine tuned it opened up all sorts of possibilities. Flight enabled insects to better evade their enemies, to hunt prey, and to seek out mates and new areas of habitat. The ability to fly long distances is not something we normally attribute to the insects, but lots of butterflies, moths, hoverflies, beetles, and many more insects undertake enormous migrations every year—borne aloft on their wings—a true wonder of nature.

The singular flying abilities of insects, such as hoverflies, relies on an extremely efficient, ingenious system that combines the brute force of the wing muscles with the elasticity of the wing and that of the thorax that houses the muscles. In the most proficient flying insects, when the muscles in the thorax contract to bring up the wings, the thorax is distorted. Muscles that join the front and back of the thorax then contract, the springy thorax goes back to its original shape and the wings pivot downward.

< The delicate hind wings of an earwig are normally out of sight—intricately folded beneath the short, tough first pair of wings.

∧ True flies owe much of their aeronautical abilities to their halteres—one of which can be seen here (circled).

Insects that can beat their wings extremely rapidly also have a special type of muscle that is unique to insects—the so-called asynchronous muscle. Normal muscles need an electrical signal from the nerves for every contraction, but asynchronous muscle can contract multiple times with each nerve signal. This allows extremely rapid wing beats. In some midges this can be more than 1,000 beats per second! Indeed, true flies, such as hoverflies and midges, are the real experts when it comes to flying. In these insects, the second pair of wings has been reduced to mere stubs—the halteres. These tiny, inconspicuous structures are crucial to the flying abilities of the true flies as they beat along with the wings and act like tiny gyroscopes. The fly uses the information from the halteres to fine tune its position in flight and precisely control the muscles that power the wings and stabilize the head.

In beetles it is the first pair of wings that have been greatly modified to form a tough, protective shield over the abdomen, called the elytra. Elytra are key to the success of beetles, since they allow the otherwise soft abdomen to be better protected, and they enable beetles to live in places where soft-bodied animals would otherwise be squashed, such as the tight spaces between tree bark.

Miniaturization

We overlook insects because they're generally tiny animals, but this small stature is another reason they are so successful. A smaller body is less "expensive" to produce and maintain, especially in terms of the various systems that are needed to get around the problems of ventilation, nutrient distribution, and excretion. Small animals can also exploit niches that are completely inaccessible to larger animals.

Insects might be small, but they are also incredibly complex. Remember that these animals have tissues, organs, and organ systems. The brain of a honey bee has around 850,000 neurons and it is capable of complex behaviors, so we mustn't equate small with simple.

Insects, like few other animals, have embraced miniaturization by squeezing enormous biological complexity into a tiny space. The champions of miniaturization have to be the staggeringly varied parasitoid wasps. These wasps are probably the most diverse of all the insects, but it's difficult to estimate just how many species there might be because they're so poorly studied. Some of them are so tiny—much smaller than some single-celled beings—that the full stop at the end of the last sentence could comfortably contain several of them. How is this possible? How can a body of tens of thousands of cells be so tiny? Inside the head of a fairy wasp there is a brain, which reaches out to the rest of the body via nerve cords and nerves. In some of the smallest wasps, this brain is composed of only 4,600 neurons, but they process the information streaming in from the senses to control complex behaviors, such as flying, walking, finding a mate, and seeking out hosts. In addition, these microscopic bodies contain muscles, a complex gut, the insect equivalent of kidneys, and lots more besides.

To become very small, these microscopic insects have simplified some of their organ systems, but a cell can only get so small until more drastic modifications are needed. One way of shrinking cells is to get rid of the nucleus. This happens in the central nervous system of these miniature marvels and allows more cells to fit in each space. The nerve fibers of these insects are so thin that they shouldn't be able to work in the normal way, and it has been suggested that their nervous system may in fact be mechanical rather than electrical.

The tiny size of these wasps enables them to exploit the smallest niches with many of them completing their development within the eggs of other insects. In the smallest fairy wasps, the eyeless, wingless males remain within the host egg and mate with all their sisters before they disperse.

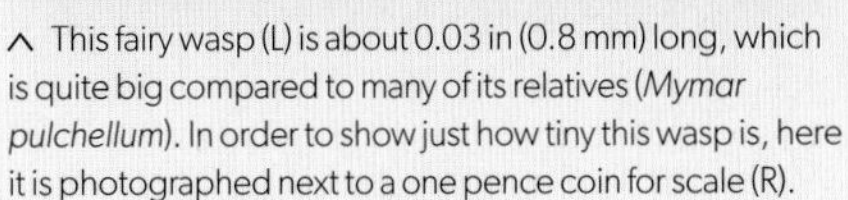

^ This fairy wasp (L) is about 0.03 in (0.8 mm) long, which is quite big compared to many of its relatives (*Mymar pulchellum*). In order to show just how tiny this wasp is, here it is photographed next to a one pence coin for scale (R).

Metamorphosis

There can be few phenomena in nature that are as marvelous as metamorphosis. To see an insect change from a larva into a pupa and finally into an adult takes some beating. When you look at a caterpillar or a maggot and then the moth, butterfly, or bluebottle fly that they become, it is difficult to grasp how these wildly different animals are related at all, let alone the same animal. The process of metamorphosis has captivated people for thousands of years and is another reason these animals are so successful. The most diverse groups of insects—the beetles, flies, wasps, bees, ants, butterflies, and moths—all go through metamorphosis. Some even go through what is known as hypermetamorphosis, where an active, hatchling larva turns into a grub-like larva that grows and pupates into the adult.

Generally, insect larvae look like soft targets. They're mainly soft-bodied and slow moving. It's true to say that a good proportion of insect larva get picked off by pathogens, parasitoids, and predators, but these shortcomings are more than compensated for by the very process of change from one form into another. Crucially, a separate larval stage and adult stage allow a division of labor in the life of an insect. The larval stage is an eating machine—dedicated solely to growth—while the adult gets all the fun and can spend its time mating and finding new areas of habitat. The other masterstroke of this strategy is that because the larva and adult are so different and typically live in different places they won't compete for resources.

The pupa was once thought to be a resting stage in the life cycle of insects, but it is anything but. The pupa's calm exterior belies an incredible amount of activity. In a series of beautifully choreographed, hormone-controlled steps, the body of the larva is dismantled and the adult form assembled.

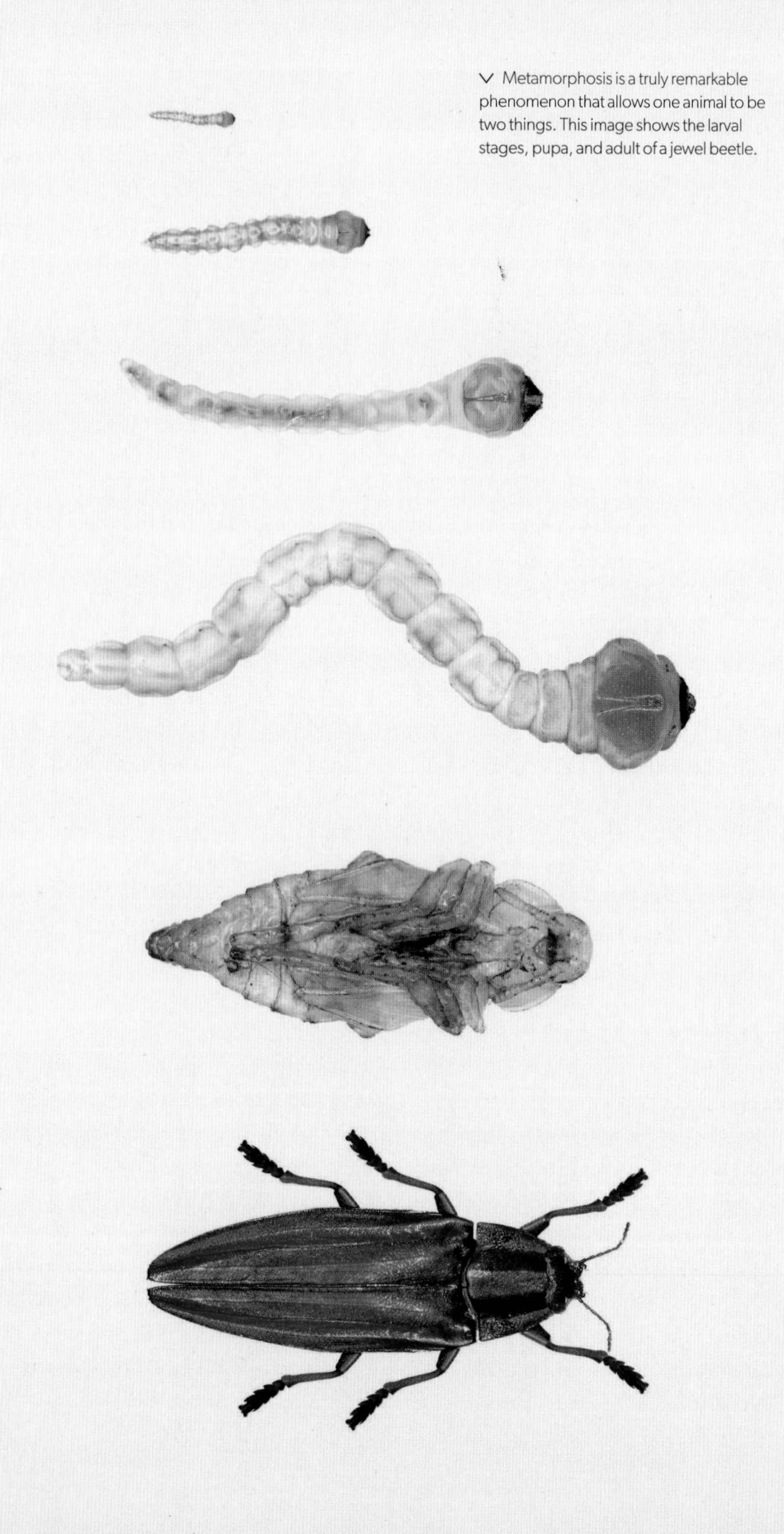

∨ Metamorphosis is a truly remarkable phenomenon that allows one animal to be two things. This image shows the larval stages, pupa, and adult of a jewel beetle.

There's still a huge amount to learn about this remarkable phenomenon. It was once thought that all the larva's tissues were broken down during pupation to create a "soup," but new research has shown this is not the case. Some of the tissue is broken down and new structures develop from clusters of cells known as imaginal discs, (for example, the muscles), but others are retained and remodeled (for example, the gut, trachea, and some parts of the nervous system). The imaginal discs can even be active before pupation. Indeed, memories formed by the larva (yes, insects have memories) are retained in the adult insect, so the connections between nerve cells must be maintained during this transformation.

Senses

The senses of insects are as refined as those of much larger animals, but their small size means that we overlook their complexity. The sensory structures of insects are often invisible to the naked eye. Look closely at an insect, perhaps through a microscope and you'll see it's furnished with an array of senses—the acuity of which contributes to their success.

There are sensory cells on the surface and within their body that sense stretching, bending, compression, and vibration. Many of these are used to sense the environment, such as the minuscule movements of air that might indicate the proximity of prey or predator, while others provide information on the position or orientation of the body. In cave-dwelling insects where eyes are of no use because of the darkness, they often have long sensory setae (sensory, hair-like structures) for detecting the movement of prey. In some beetles and true bugs, sensory cells have even been modified to detect infrared. The offspring of these insects can only develop in the wood of burnt trees, and their remarkable heat sensors enable them to find freshly charred trees.

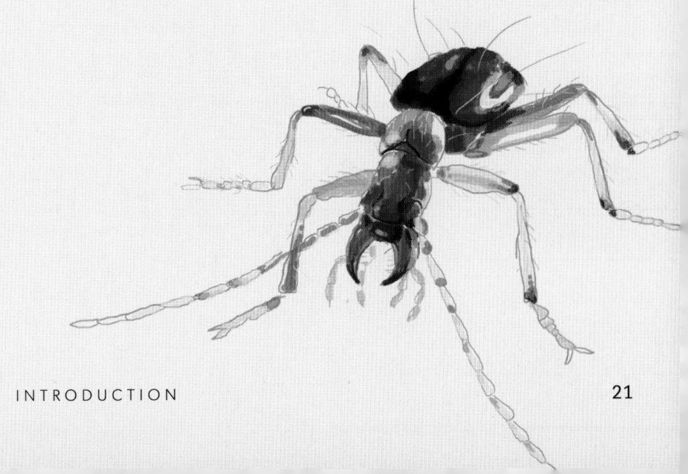

> Insects have sensory "hairs" on their body. These are at their most elaborate in eyeless cave insects, such as this beetle, *Arctaphaenops muellneri*.

Not all insects can detect sounds in the way that mammals do, i.e. by picking up movements of the air via a drum-like membrane. Receptors that pick up the tiniest vibrations traveling through the substrate are common. Some insects do have organs with a drum-like membrane and they provide some of the sharpest hearing in the animal world. In some true bugs these ears are located on the thorax. Grasshoppers, cicadas, and moths have ears on their abdomen, whereas you must look to the front legs to find the ears of crickets and katydids.

The most sensitive known ears among the insects belong to a parasitoid fly (*Ormia ochracea*) that has to pinpoint the location of its hosts—crickets—by their song. Depending on where the sound of the singing male cricket is emanating from, the tiny ear drums will reverberate at slightly different times. This difference may be as little as 50 billionths of a second, but it is enough to allow the fly to directly home in on a singing male cricket. It doesn't have to stop and cup its ears; it just precisely identifies the source of the sound. Even if the cricket stops singing, mid-homing, the fly can approximate its position from the last sound it made.

The ability to detect chemicals, i.e. taste and smell, is extremely acute in insects. You won't find a nose on an insect, well not one that you see on the face of a mammal, but they are bristling with all manner of receptors for detecting chemicals. These are normally concentrated on the mouthparts, but they can also be found on the antennae of some insects, as well as the feet and ovipositor of others. The life cycle of many insects hinges on being able to detect mates and food from afar, so their ability to sense individual molecules in the air is remarkable. Flies and beetles that need decaying animal remains for their larvae can detect the chemical signatures of death from many miles away. The adult lives of insects are often very fleeting (sometimes a couple of hours) and the window in which to detect a receptive mate is tiny, so males must be able to detect the

< The antennae of this male glowworm are able to detect tiny quantities of female pheromone to lead him to a mate (*Calyptocephalus* sp.)

> Many insects have very sensitive ears. Bush cricket's ears are on their front legs (circled).

merest whiff of a female on the breeze. For this reason, male insects often have elaborate antennae for just this job. The ornate, often branching antennae of male insects are effectively molecular sieves able to screen individual molecules from the air. By following the increasing concentration of these molecules, the male will eventually be led to a potential mate, unless he's beaten to it.

Insects have very snazzy eyes. Often, they have more than one type of eye—compound eyes and so-called simple eyes. Simple eyes aren't exactly simple. They're exquisite, tiny structures for sensing light, and one type—the lateral ocellus—can probably resolve outlines of nearby objects. The other type of simple eye—the dorsal ocellus—is only ever found with compound eyes and it is thought they may be important in tweaking the sensitivity of the compound eyes to light intensity.

Compound eyes can be relatively enormous and in some insects the head is little more than eye. Every compound eye is made of individual, tightly packed units called ommatidia. Each ommatidium has a lens and associated cells that detect light, so each ommatidium captures an image that is relayed to the brain. Up until recently it was assumed that this arrangement could only provide relatively low-resolution images. However, the photoreceptor cells underneath the lenses of the compound eye move rapidly and automatically in and out of focus, which

^ Owlfly compound eyes are split into two parts. The top part is exclusively UV sensitive. The lower part has more sensitivity in the blue-green wavelength range. This probably helps them to pick out prey against the sky.

provides a view of the world that is much sharper than previously thought. As well as giving insects a sharp view of the world, these compound eyes also have a very wide angle of view and are second to none when it comes to detecting movement. Many insects can also detect light that is invisible to humans.

Reproductive Potential

Phrases such as "to breed like flies" do have a basis. Indeed, insects generally are synonymous with fecundity. Many female insects lay lots of eggs, often hundreds. Insects such as oil beetles, which have strange, convoluted life cycles, must produce thousands of eggs to off-set the low chances of any one offspring reaching adulthood. Some of the supreme egg layers though are the queen ants and termites. A termite queen may produce more than ten million offspring in her long life, but even this is meager compared with some leafcutter ant queens who can produce 150 million young in their lifetime.

Most of the eggs laid by a female insect will hatch, and in several types of insect the generation time is very short. Aphids in particular are renowned for the short generation times and in some species, this is as little as five days. Unbound and in perfect conditions the populations of the most fecund insects can explode. Large numbers of eggs and short generation times are not the only secrets to the reproductive success of these animals. Female insects can store sperm from a single mating and make it last a lifetime—enough to fertilize all the eggs they will ever produce. In any given population of an insect, females often outnumber the males, as a few of the latter can more than meet the sperm demands of the females. In aphids, thrips, stick insects and many other insects, this trend has reached its ultimate conclusion, as males only feature in part of the life cycle, are very rare or have been erased completely leaving parthenogenetic females pumping out clones of themselves. This is how mass gatherings of aphids can appear on plants, seemingly out of nowhere.

All these factors combined mean that, as a rule, insects are very good at making more insects. Perhaps the most important aspect of this rampant reproduction is that it churns out mutations, a tiny proportion of which will be beneficial and allow the owners to adapt to a continually changing world. This is perhaps most easily understood when we think about insecticide resistance in the insects that we want to get rid of. When we douse the environment with a new insecticide the initial results are dramatic. The targets are seemingly vanquished, but there will

∨ Without the need for sexual reproduction for part of their life cycle, aphids can establish enormous populations with alarming speed.

< The ability of a caddisfly larva to construct this case from silk, snail shells, and plant debris is completely innate.

^ A female spider-hunting wasp with her prey. The ability to find and dispatch prey in a very specific way and provision a nest for her offspring is hard wired in her brain (*Auplopus carbonarius*).

always be some survivors, a small proportion of individuals that have a chance mutation that renders them immune to the insecticide. These resistant individuals go on to breed, passing this resistance to their offspring. In a relatively short amount of time, all the insects in a population will be resistant to the insecticide. This is evolution in action and the same process applies to every aspect of an insect's life. There will always be the genetic resources out there that allow adaptation to the challenges that life throws their way.

Complex Behaviors

The behavioral repertoire of insects is immense. From the intricacies of their life cycles, through to the finding of food and mates, and the evasion of their many enemies, insects do some remarkable things. Much of what they do is innate, in other words the behavior we see is encoded in their DNA. This includes very elaborate actions. When we watch a caddisfly making its remarkable case or a hunting wasp diligently snipping the legs from its spider prey before it transports the victim back to its nest, it's hard to believe that these complex actions aren't learned. The truth is that the caddisfly and the wasp don't need to learn these things—they're hard wired—somehow this knowledge is encoded in their DNA.

Although innate behaviors account for much of what we see insects doing, some of them can change their behavior because of experience. In other words, they can learn. As an example of how elaborate this learning can be you only need to look at honey bees. When returning from a successful foraging sortie, a worker honey bee will do a strange dance—the waggle dance. This has been known about for a long time, probably for as long as people have been keeping bees, but it took the genius of the ethologist Karl von Frisch (1886–1982) to figure out what it meant. Far from being a celebratory jig, this "dance" is the worker using symbolic language to teach her sisters about the location of food, water, and new nesting spots. The fact that these small animals can memorize and relay this information to others of their kind to learn is something to marvel at.

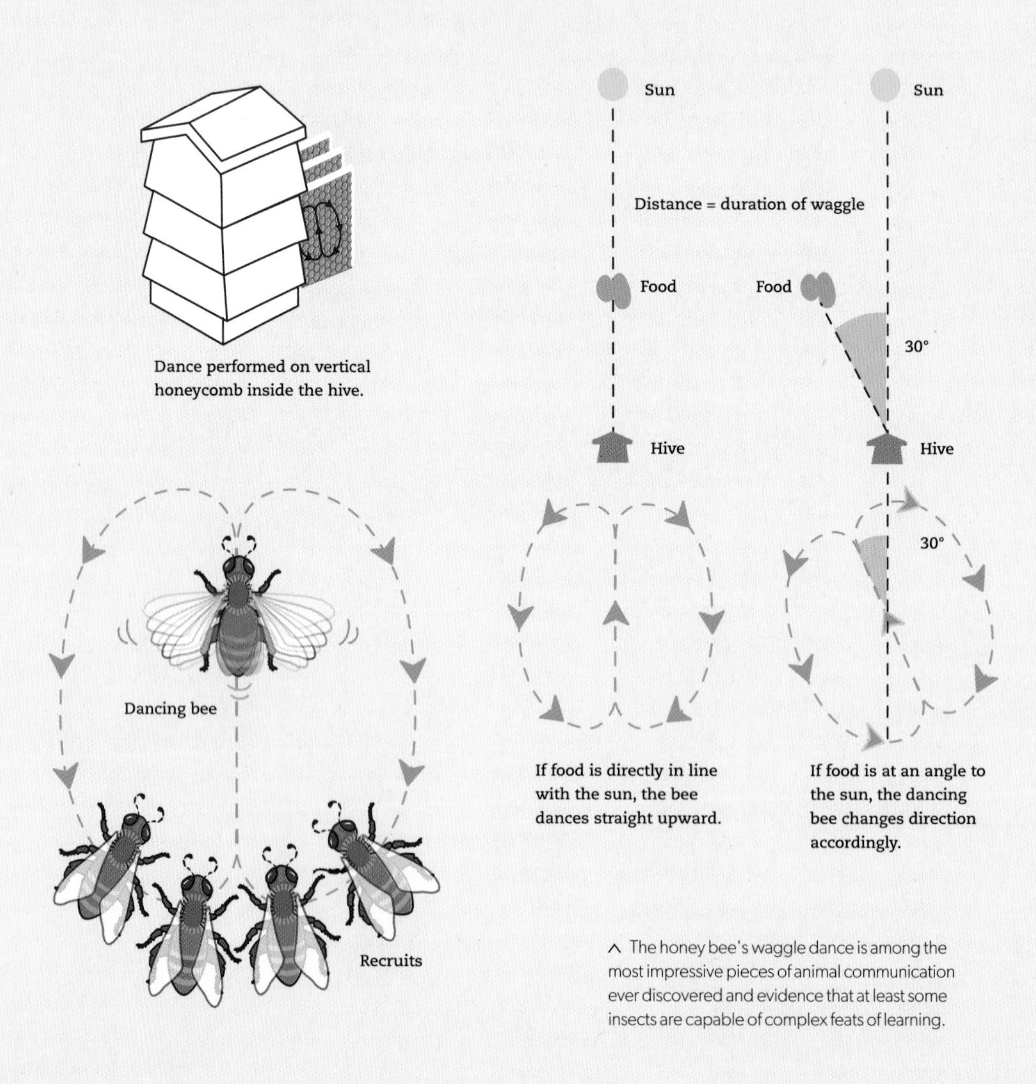

∧ The honey bee's waggle dance is among the most impressive pieces of animal communication ever discovered and evidence that at least some insects are capable of complex feats of learning.

Other more recent studies on insects have found that bees can count and that they can discern whether two symbols are the same or different. Not only that, but social wasps have also been shown to recognize the faces of other wasps. These examples show that some insects are capable of impressive cognitive feats.

There's still an awful lot to discover about the learning ability of insects and we've only just scratched the surface. Of the more than one million species of insects that have been described, only a handful have been studied to test whether they are capable of learning. Most of the insights so far come from social insects and it's not really a surprise these animals are capable of learning because they live in large, complex groups where there's ceaseless interaction between individuals conveying information about many things, such as food and threats.

About this Book

In this book we explore the lives of insects. Within the confines of 40,000 words I had to be very selective in what examples I used in each chapter, and I have focused on the bizarre and remarkable. Bear in mind that the insects are an enormous group of animals, with many more species than all the other animal groups combined. Even though only a small proportion of insect species are well studied, these "known" ways of life are still stunningly diverse. Even among the well-known species there are still discoveries to be made and just think about all the other insect species out there. The ones that have been described by taxonomists, but the lives of which are a mystery, and the millions of species that are still to be collected and described. It would take an army of biologists thousands of years to understand exactly how all these insects live. You can reflect on all those species, the ways in which they might live, and the web of interactions they have with other living things. The complexity is mind-bending.

The facts in this book about how certain species live and why they do the things they do, were gleaned by patient observation, often over many years and sometimes over whole careers. The curiosity, patience, and dedication of these naturalists and scientists is sometimes as remarkable as the insights to which they led. The drive to ask questions, and understand more about life on Earth, is what makes us who we are, and it's something we should all celebrate and nurture. The great thing is that anyone can help to fill in these blanks—there's enormous scope for exploration and discovery within entomology. Watching and studying insects can take you to some amazing places, but equally, discoveries can be made in your own backyard. All you need to do is get out there and look. Hopefully, this book will give you a taste of the remarkable lives of insects, help you to make sense of some of the things you might see if you watch them, and encourage you to look more closely at these endlessly fascinating animals.

1

LIFE CYCLES

Insects, like every other animal, have a life cycle—normally starting with an egg and going through various stages to the adult. This journey from egg to adult is full of surprises.

From aphids, which alternate between sexual reproduction and cloning themselves to the fiendishly complex life cycle of the telephone-pole beetles and bee flies. When we look closer, we can find exquisite diversity in how insects grow and develop.

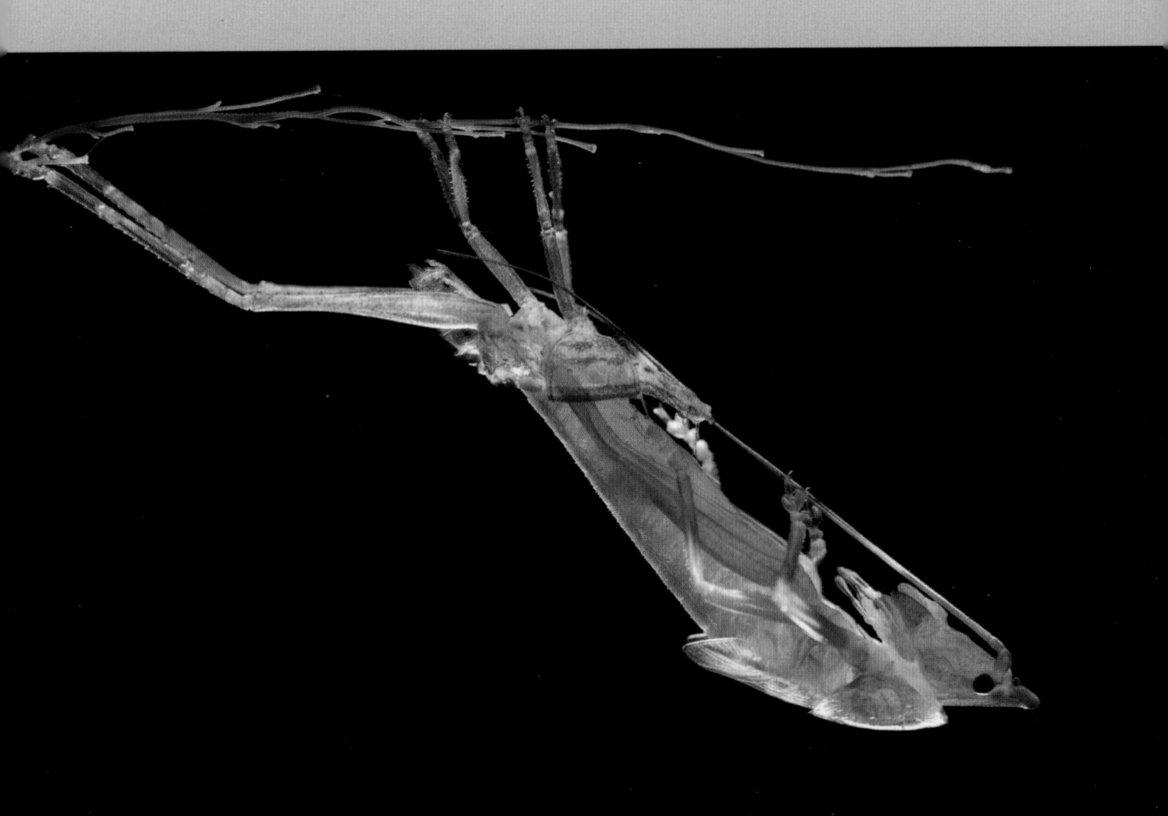

FROM EGG TO ADULT

In most species an insect starts out as an egg—often very ornate when you see them magnified. They range in size from the microscopic, 0.0008 inches (0.02 mm) capsules of some parasitoid wasps to more than 0.4 inches (10 mm) in some bush-crickets, bees, and beetles. Female insects will try and lay their eggs in a place that gives their young the best start in life.

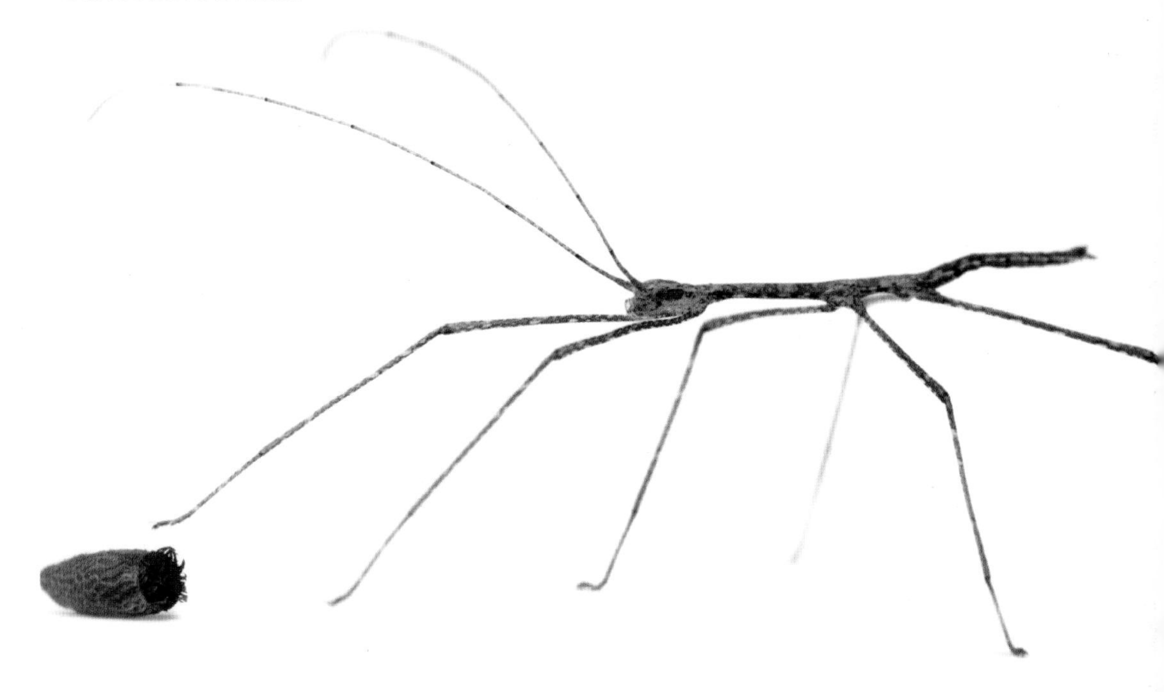

∧ The large eggs of stick insects often mimic seeds.

< Bush cricket shedding its old exoskeleton in order to grow.

In those insects that don't go through metamorphosis, the egg hatches into a nymph, essentially a miniature version of the adult, and in those that do, it hatches into a larva. The larvae or nymphs grow in stages, dictated by the limits of the exoskeleton. The nymphs carry on growing like this until they reach the adult stage. Mature larvae have to pupate to make the transition to the wildly different adult form.

^ The most challenging few hours of a dragonfly's life—when it turns from an aquatic animal into a creature of the air.

Growing in Stages

Sometimes you might find an insect that looks unusually pale. These are often insects that have just shed their exoskeleton. The new exoskeleton underneath is not fully hardened—a process that can take some time. In this teneral state they're very vulnerable. When an adult insect emerges from a pupa or from the final nymph stage, the wings are tiny and crumpled. They must be inflated and hardened before they can be used for flight. Hemolymph is pumped into the wings and they slowly grow to normal size. The insect needs space to do this and anything pressing on the extremely delicate wings will stop them from inflating properly and they'll be useless.

FIG WASPS
Agaonidae

Most people will not have seen a fig wasp, but these tiny insects have a remarkable life cycle, most of which happens within the confines of a fig and the details were only pieced together by some serious sleuthing. A fig is really a cluster of flowers that has grown in on itself. At one end of the fig is a tiny hole that connects to its center via a narrow tunnel. When the fig is ready to be pollinated a female fig wasp crawls through this tunnel, which is such a tight fit that she loses her wings and antennae in the process. Once in the central cavity she busily lays eggs into some of the fig flowers, distributing pollen between the flowers. Her passage into the fig was a one-way ticket, but her young develop and eventually emerge as adult wasps. The male wasps have no place in the outside world. Instead, they are wingless and equipped with powerful jaws and limbs to battle their brothers for the right to mate with their sisters. The victorious males will mate with their sisters and chew a hole in the wall of the fig through which their sisters will escape to fly off and find a fig to call their own, carrying some pollen from their birth fig.

DIVERSITY AND IMPORTANCE

- 750 fig species
- 10,000 species of fig wasps and their parasitoid wasps
- A huge variety of large forest animals depend on figs for food

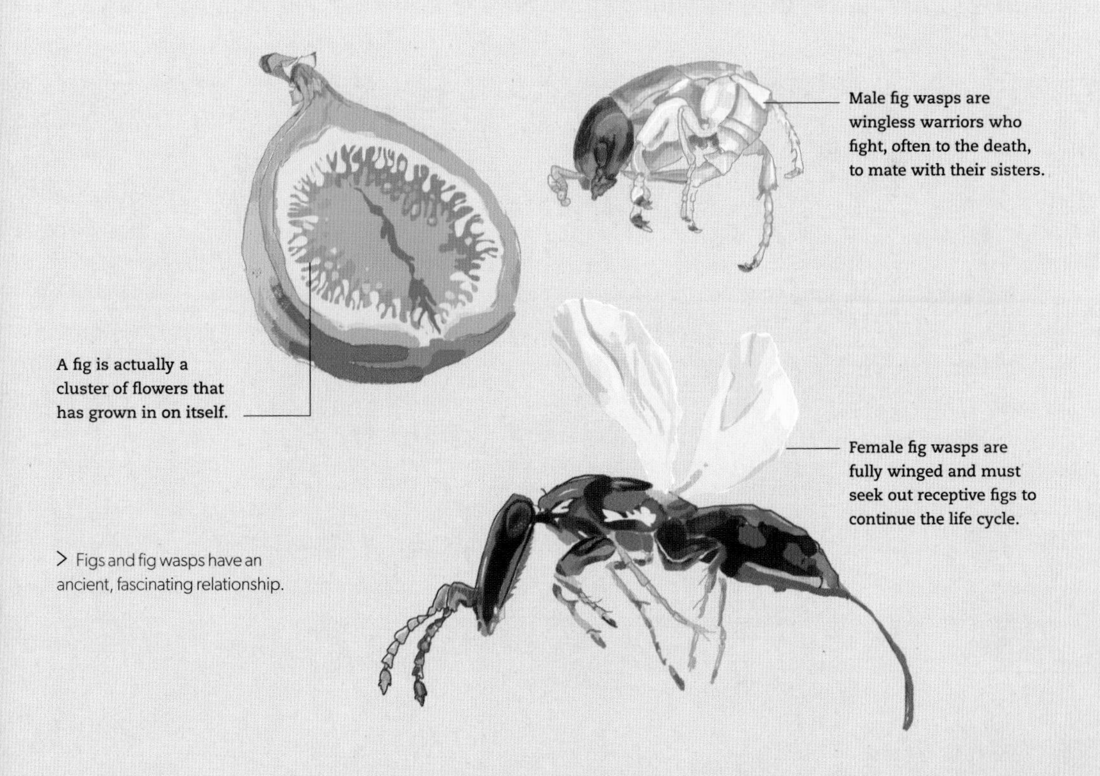

> Figs and fig wasps have an ancient, fascinating relationship.

Unusual Life Cycles

Within this standard path from egg to adult there is enormous variation and many insects have evolved some really mind-bending life cycles. Oil beetles (*Meloidae*) lay thousands of eggs, often in a small burrow in a flowery place where unsuspecting bees might be foraging. The eggs hatch not into normal, grub-like larvae, but small, very mobile beasts called triungulins. These scuttle up the nearest flowers and wait for a bee to land in search of nectar and pollen. They climb aboard the bee and get carried back to its nest where they proceed to destroy its eggs. The triungulins change into grub-like larvae that make short work of the provisions the female left in the nest for her own young. The beetle larvae grow fat, pupate into new adult oil beetles. Some oil beetle triungulins have added another level of complexity and cooperate to mimic the shape and odor of a female bee to attract a lustful male bee. When a male bee attempts to mate with this writhing cluster of young parasites, they clamber aboard and cling to his body and wait for him to find a real female to mate with so they can get into a nest.

The pleasing, fluffy flies known as bee flies have adopted a similar strategy to the oil beetles in that the egg hatches into a mobile triungulin, but their mode of getting the triungulins to the right place is more direct. The female bee fly buzzes her back-end against dry soil to give her sticky eggs a bit of ballast. Next, she hovers over the nest holes of her hosts—solitary bees—and dive-bombs her dirt-coated eggs into the burrows. When the bee fly triungulins hatch they plunder the host's nest in the same way as the oil beetles.

∨ The cuddly appearance of an adult bee fly belies the complexities and grisly nature of its life cycle.

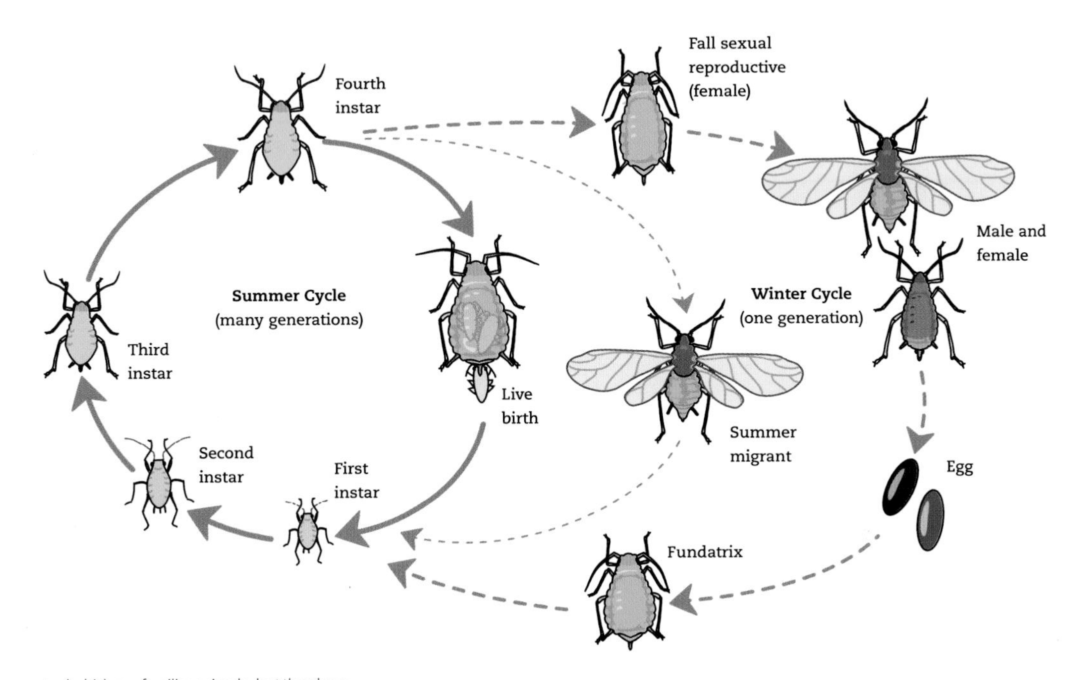

∧ Aphids are familiar animals, but they have some of the most remarkable life cycles of any insect.

Triungulins hitching rides on bees to plunder their nests is pretty good, but this is nothing compared to the telephone-pole beetles (*Micromalthus debilis*). If you've never heard of these beetles, then you've been missing out. Its life cycle is about as extreme as it gets. No (or very rare) sex. Just weirdness. Most populations of this wood-munching beetle consist entirely of female larvae that give birth to triungulin larvae. The triungulin molts and turns into another limbless female larva that gives birth to yet more triungulin larvae. Sometimes, one of these legless female larvae pupates to give rise to the adult female. The weirdest part is when one of these limbless larvae gives birth to a male egg. The male larva hatches, shoves his head up his mum's genitals and proceeds to eat her from the inside out. After this matricide, the male larva pupates and turns into an adult male. This doesn't happen very often, so the males are extremely rare.

Aphids are among the most prodigious reproducers among the insects, thanks to their ability to dispense with sex for much of their life cycle. During the warmer weather, aphid populations can explode on their favored food plants because they don't need males, which are a waste of space in many ways. The females you can see on a plant might be clones from a single female who alighted on that plant a few days before. Take a close look at one of these aphid aggregations and you'll see new clones being born. Within each of these new clones is another developing clone and yet another one within her.

COURTSHIP

Before two keen insects can mate they must find each other and this is no mean feat for these small, generally short-lived animals. Not only that, but most insect species are rare, which adds to the challenge of finding a mate. Fortunately, they've evolved all sorts of tricks to solve these problems, including odors, songs, and light shows.

Senses of smell and taste are very refined in insects and they are used to detect pheromones given off by members of the opposite sex. This is extremely effective and us humans have tapped into this odor channel to trap insects we don't like by using synthetic pheromone lures.

Lots of insects also sing to attract mates. Forests, grasslands, and every habitat in between are alive with a cacophony of strange sounds, much of which is the work of insects. Tropical forests in the daytime buzz with the songs of cicadas. To be fair, cicada songs are not very musical to our ears at least, but they do generate an impressive din by rapidly vibrating

∧ Male Grant's stag beetles fight each other using their enormous mandibles.

> By rubbing its penis against some abdominal ridges, the lesser water boatman can produce a 99.2-decibel song.

^ Lots of beetles use light for the purposes of courtship. Here, a long exposure photograph shows lots of these beetles in action.

membranes on their abdomen. Grasshoppers and crickets sing by stridulating—rubbing two body parts together—leg against wing in the grasshoppers and wing against wing in the crickets. Some ground-dwelling crickets reach a bigger audience by singing from a burrow with the acoustic qualities of a trumpet. For maximum volume, the lesser water boatman "sings" by rubbing its penis along some ridges on the underside of its abdomen, generating a 99.2 decibel. This is quite a feat for an animal that is only 0.08 inches (2 mm) long! Sadly, this song is lost on us, as the volume falls hugely when the sound moves from water to air.

Light has also been harnessed for the attraction of mates. In beetles such as glowworms, the wingless larva-like female glows and attracts the attention of males flying nearby. They drop to the ground to try their chances. In other light-show beetles, both the female and male glow.

Sex results in competition for mates and the best breeding spots. These pressures explain some of the weird and wonderful forms and behaviors we see among the insects. Take the stalk-eyed flies, the enormous mandibles of Darwin's beetle, the strange appendages of the tiger moth *Creatonotos gangis*, the gigantic sperm of *Drosophila bifurca*, and the feathery legs and inflatable abdominal sacs of some dance flies. All of these adaptations are the product of sexual selection—the struggle to reproduce. In many insects, the males engage in combat to get access to mates, while others simply compare their outlandish ornamentation to deem who's the fittest.

In those insects where the biggest males with the biggest weapons normally get the females there has sometimes been a parallel evolution of weedy, sneaky males. The size and ornamentation of the butch males comes at quite a cost and if a small male can succeed in passing on his genes without being big or having outlandish ornaments, then he will. One of the nicest examples of this is the rove beetle, *Leistotrophus versicolor* (see overleaf).

THE ROVE BEETLE
Leistotrophus versicolor

This beetle lives in the rain forests of Central and South America, where, like lots of other rove beetles, it makes a living by seeking out decaying plant and animal matter in order to feed on the adult insects and larvae that make use of these ephemeral resources. These honeypots don't last long in the supercharged biological activity of the hot and humid tropical forests, so when normal male rove beetles find these resources they guard them because they also attract females, allowing a male to assemble a harem.

Males of this rove beetle are divided into two types: big males and small, effeminate ones. The small, effeminate males can find honeypots, but they have little hope of defending them against the bigger males, so their chances of building a harem are small. These males have evolved another means of making sure they pass their genes onto the next generation. They sneak past the big males using their effeminate appearance as a disguise and under the harem owner's nose they mate with the females he has been so carefully guarding. This strategy is almost flawless, but now and again the effeminate male is caught in the harem by the harem owner and the only way he can avoid being torn limb from limb is by assuring the aggressor of his femininity and submit to a "mating." Even after this, the effeminate male carries on sneakily mating with the females in the harem, but perhaps more cautiously.

∨ Sneaky, effeminate males pass on their genes without going to the trouble of guarding a harem.

Ritual Combat

Ritual combat among male insects has certainly driven the evolution of some fabulous forms, but this is also true of males that simply compare their oddities to see who's the best. In some fungus weevils, the males defend females and size up to other males who try their luck, but instead of scrapping they just compare their stalks, eye stalks that is. The males with the more developed eye stalks win these disputes.

∧ A variety of insects size up opponents by comparing eye width. This is the fungus weevil *Exechesops leucopis*.

∨ Lots of male beetles have horns or enlarged mandibles for fighting other males.

To see this trait taken to the extreme we must look to the flies. Eyes on stalks have evolved a few times in the flies, but the champion has to be *Achias rothschildi* from that island of wonders—New Guinea.

INSECT GATHERINGS

Gatherings of insects can often be seen seemingly dancing and signaling to one another on patches of ephemeral resources, such as dung, rotting fruit, and carrion. It's also common to see small swarms of insects hovering above bushes or prominent high points in the landscape, as well as individual male insects perched on these same high points. These conspicuous behaviors are typically all about reproduction and are known as lekking. Males of the insects in question will gather and females will come along to find a mate. Less commonly, the roles are reversed with males coming to inspect the females. There's a lot of communication going on in these gatherings, only a fraction of which has been decoded. Pheromones are certainly important as are visual signals, such as bold colors and patterns, and inflatable sacs. Wing beat frequency is another way in which the insects signal what a good catch they are, and it also seems that the flash frequency of light reflected off the beating wings is important in some species.

∨ Lots of insects, especially aquatic species emerge en masse to increase the chances of finding a mate and to swamp predators.

∧ A nuptial gift. This female cockroach (left) is lapping at a secretion from glands on the back end of the male.

< The chemical defenses of the female bella moth come from the male's spermatophore (*Utetheisa ornatrix*).

Nuptial Gifts

To grease the wheels of courtship, some insects come bearing gifts. These range from tasty salivary secretions or edible matter regurgitated by the male, the oozings from antennal secretions, anal droplets, some toxins, a nice prey item or, most delectable of all, spermatophores, which are nutritious blobs of gloop surrounding the male's sperm. All of these gifts help to give the female a burst of nutrients to help mature her eggs and while she's preoccupied with it the male can take his time with copulation, giving his sperm a better chance of fertilizing the eggs before those of rivals.

The spermatophore of some insects is often complex, and can consist of a tough outer cover, nutritious material, and sperm. In the humble small white butterfly (*Pieris rapae*), the offering of the male accounts for 13 percent of his body weight. The outer case of this butterfly's spermatophore is so tough that the female has a set of jaws in her reproductive tract to chomp through it to get at the nutritious material within and to free the sperm. Her reproductive tract also has to function like a stomach to digest the nutritious goodies. Even more bizarre is the fact that this spermatophore is constructed inside the female by the business end of the male's genitals during mating.

The spermatophore of the bella moth (*Utetheisa ornatrix*) has added kick as it also contains defensive toxins. During the final stages of mating these spread through the body of the female and into her ovaries conferring her and her offspring some protection from predators.

There is much still to learn about these nuptial gifts, and the secretions that are passed to the female during mating. It seems as though some secretions from the male can affect the female in lots of different ways, such as altering her lifespan, reproduction, and feeding behavior to increase the chance of the gift bearer passing on his genes. The ultimate nuptial gift is where the male insect offers up bits of his own body or even his own life to maximize his chance of passing on his genes to the next generation, although this behavior can also evolve as a way of the female controlling which males will father her offspring.

Normally, it is male insects that try to impress females with nuptial gifts, but there are cases where the roles are reversed. Take the Mormon cricket (*Anabrus simplex*), his spermatophore is so enormous—about 30 percent of his bodyweight—that it is coveted by females and they come flocking after he stridulates for a few minutes in the morning. The females come to blows over who will get access to the male and his massive spermatophore. Even the victorious female may be rejected by the male if he deems her insufficiently weighty when she climbs on his back to mate.

The roles have also been reversed in Zeus bugs (*Phoreticovelia disparata*) and it is the female who offers a gift. The small male rides on the female's back and sucks up a waxy substance she secretes. This strange arrangement frees the female from fighting off unwanted male attention.

∧ A female Mormon cricket with a spermatophore (surrounded by its gelatinous coat) that she has obtained from a male (*Anabrus simplex*).

< Oil beetle sex also involves the male-to-female exchange of defensive chemicals to grease the wheels of reproduction (*Meloe* sp.).

MATING

Mating in insects can be a very brief, all-over-in-a-split-second experience, or it can last for many hours, even days—the record is 79 days, set by stick insects. Some male insects have evolved all sorts of guarding tactics before and after mating, to try and maximize their chances of fathering the female's offspring.

The tandem flying behavior of dragonflies and damselflies is one such strategy. The males of these insects have a pair of special claspers at the tip of their abdomen for grasping hold of the female. Sperm from different males will compete within the reproductive tract of a female, so male dragonflies and damselflies have a spoon-shaped penis for scooping out the sperm of rivals who have previously mated with a female.

The urge to mate with a virgin female is so strong that it drives some male insects into a frenzy. A freshly emerged female solitary bee probably wishes she hadn't bothered when she quickly becomes the center of a scrambling ball of hormone-addled males all desperate to mate with her. As soon as she acquires a mate, the clamoring horde quickly goes its own way to try and find another virgin female. Male thynnine wasps literally sweep the female off her feet and away from competing males. In these wasps, the females are flightless and much smaller than the male. A female ready to mate attracts males by wafting pheromones from glands on her head and before long one swoops down and flies off with her to find a quiet spot

∨ Role reversal. *Neotrogla* cave bark-lice females have a penis. She mounts the male (L) and uses the penis to keep them locked together (red, R). Roles have flipped here because food is scarce in their dry caves and it is the females who want to mate as much as possible because of the nutritious spermatophores produced by the males.

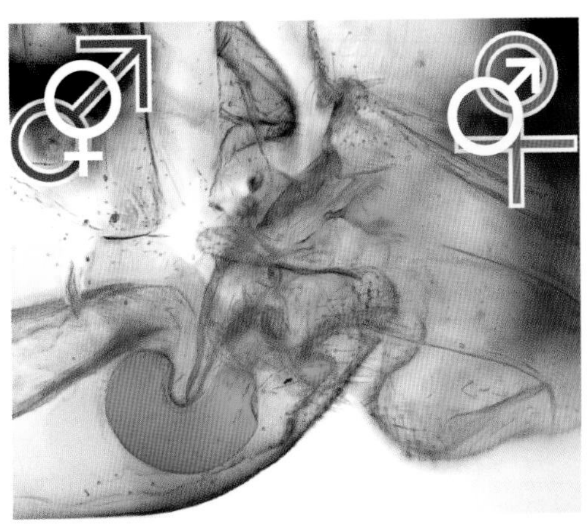

for a lengthy bout of mating. Remarkably, some orchid flowers mimic the odor and appearance of these female wasps, duping a hapless male into a pointless mating during which he pollinates the orchid.

When it comes to mating, male and female insects often want different things. A male insect packs lots of sperm, which are biologically cheap to produce. In essence, he stands to gain from mating with lots of females. On the other hand, a female often has to invest a lot more time and resources in the production of her eggs, so she is choosier about whom she mates with and seeks quality over quantity. This dichotomy can be so extreme that it actually drives an evolutionary arms race between the sexes.

One of the most dramatic examples of this and the best studied is in the much-maligned bed bug (*Cimex* spp.). A female bed bug can snub an amorous male if she's not interested in mating by curling her abdomen underneath. This behavior drove the evolution of a counter behavior, known as traumatic insemination, a name which gets across some of the brutality of what happens. The snubbed male simply stabs the female's abdomen with his penis and deposits his semen into her body cavity where the sperm make a bee-line for the ovaries. When male bed bugs first started doing this long ago in their evolutionary history it was bad news for the

∨ In many insects, males can detect the imminent emergence of a female. This male ladybird is guarding a female pupa in order to mate when she emerges (*Illeis koebelei*).

^ Dirty tricks. The male red postman butterfly taints the female with pheromones that make her smell bad to rival males (*Heliconius erato*).

females as their abdomen was ruptured, which could have led to disease and death. However, over time, the females have evolved their own counter to the penis-wielding males in the form of a special structure that receives the stabbing penis and limits the damage to her abdomen.

In other insects, the sexual conflict is more subtle. For example, in some butterflies, such as the red postman (*Heliconius erato*) and the small white (*Pieris rapae*), the male transfers pheromones to the female during mating that make her smell bad to rival males, so they give her a wide berth. Some male insects, for example the clearwing swallowtail butterfly (*Cressida cressida*), simply bung up the female's reproductive tract with a gummy plug that can be half as long as the female's abdomen. The males of the vinegar (or fruit) fly (*Drosophila melanogaster*) are altogether more dastardly. Not only do they go for traumatic insemination, but they employ toxic semen. In his semen there is a cocktail of accessory gland proteins that really mess up the female. These proteins induce the female to lay her eggs before she can mate with other males, reduce the likelihood of her remating and, as a final insult, reduces her lifespan. Where these evolutionary arms races will go is anyone's guess, but they have yielded some truly remarkable adaptations.

< Top: Female pot beetles delicately encase each of their eggs in fecal plates (*Cryptocephalus aureolus*).

Bottom: The eggs of lacewings and their relatives sit atop a stalk to give them some defense against predators.

∧ Completed egg case of the female pot beetle.

> An egg-case (ootheca) of an Amazonian mantis (*Liturgusidae*). Even with this protection, all the eggs in this case fell prey to a parasitoid wasp.

Egg Laying

As with everything else about insect reproduction, their eggs and exactly how they lay them is fabulously diverse. For example, a mosquito lays a little raft of eggs, so the larvae can take to the water straight away when they hatch. Some insects, such as mantises and cockroaches, secrete a tough case around their eggs, called an ootheca.

Stick insect (*Phasmatodea*) eggs look like plant seeds, which dupes ants into burying them and providing protection from some parasitoid wasps. In some stick insects, the seed-like eggs are even gobbled up and dispersed by birds and the tough covering of the eggs prevents them from being digested in the bird's gut.

Lacewing (e.g. *Chrysopidae*) eggs are laid atop a long thin stalk, which prevents them from being detected by ants and other predators. Some beetles encase their eggs in a neat case formed from fecal plates before flicking them into the leaf litter. In some cases, these are also picked up by ants and carried into their nests.

The infamous human bot fly (*Dermatobia hominis*) runs the risk of getting unceremoniously splatted if it tries to lay eggs on its hosts, so it gets other flies to do its dirty work. The female bot fly catches another fly, glues her eggs to it and lets this unwitting courier take them to an unfortunate host mammal. Female bee-grabber flies (*Conopidae*) have an abdomen tip a bit like a can opener in order to insert their egg into the abdomen of a living bee in a lightning-fast strike.

Some of the most sophisticated egg layers are the parasitoid wasps that have to use an almost impossibly long ovipositor to drill down to their hosts, as many insect larvae live deep within wood. The female wasp uses odors and a form of echolocation to pinpoint the position of her quarry before she drills down to it and deposits an egg.

PARENTAL CARE

On the whole, insects aren't really known for their parental care. Most just lay their eggs and move on without even giving their young a backward glance. There are some welcome exceptions to this rule though and some insects make dedicated parents.

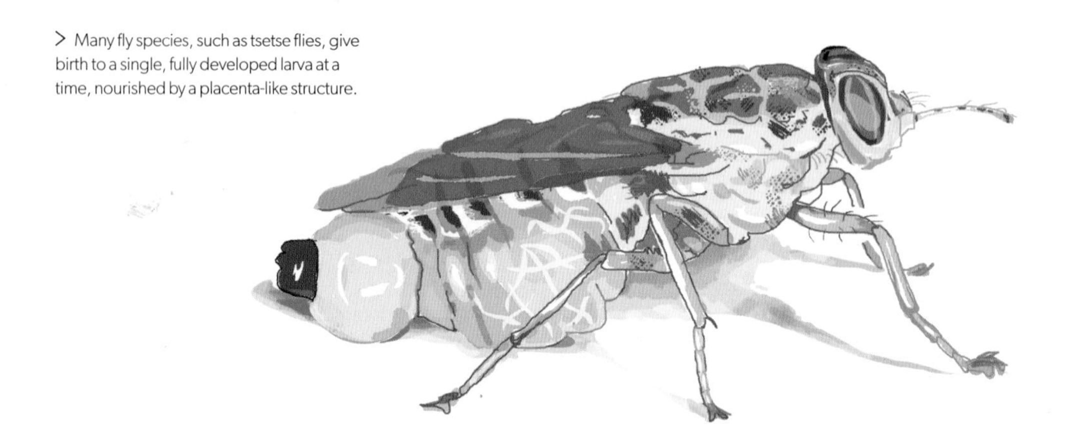

> Many fly species, such as tsetse flies, give birth to a single, fully developed larva at a time, nourished by a placenta-like structure.

There are a few insects that do not lay eggs, instead they nurture the young inside their body until they're well-formed. Tsetse flies and their relatives don't have all that many admirers, but their reproductive biology is engrossing. The females of these flies nourish a single larva at a time in their uterus where it feeds on a milky secretion produced by a specialized gland. The larva grows within the safety of its mother's uterus for as long as thirty days and can pupate immediately after it is born.

Earwigs are some of the most familiar insects and they take extra special care of their eggs, diligently cleaning them to give them the best chance of hatching successfully. Even after the young hatch they stay in the nest for a good while being tended by their mother and this childcare even continues when the young start to make forays above ground. In giant water bugs, it's the dad that tends to the eggs, brooding them on his back after the female has laid them there until they hatch.

Leaf-rolling weevils (*Attelabidae*) fashion perfectly formed little packages from leaves in which to lay their eggs. The beetles make precise cuts in the leaf to ensure that it can be rolled up in just the right way. This is a time-consuming process for the weevil, but it gives the offspring a supply of food and some protection from parasitoid wasps.

Some tortoise beetles (*Cassidinae*) diligently guard their eggs and larvae. In some species, the female sits tight over her eggs until they

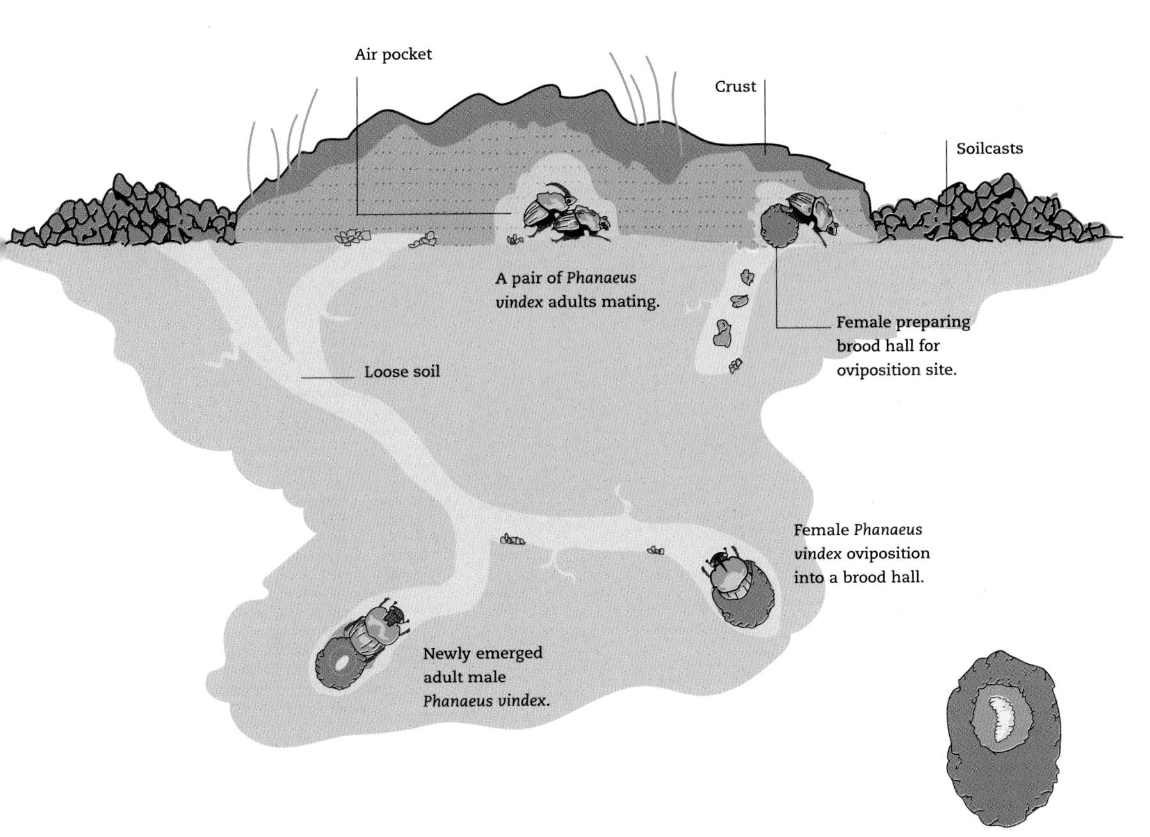

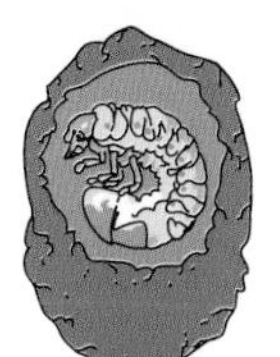

hatch and stays with her young as they grow until she sits atop a veritable mound of offspring. The larvae all face inward, offering a delightful fecal shield to predators and parasitoids that venture too close. Some shield bugs have evolved a similar strategy, standing guard over their eggs and throwing some amusing shapes if they feel threatened.

∧ Some dung beetles burrow directly beneath mammal dung to excavate their brood chambers.

> Within each ball of dung, a beetle larva develops quickly. The mother often stays with her developing brood.

Bledius are charismatic little beetles of charming habitats, such as estuary mud. The females make a tiny, wine-bottled shaped burrow in which they lay their eggs. When the tide comes in, the narrow neck of the burrow prevents the burrow from being inundated immediately, buying the female time to make a mud stopper for her little nest. After the eggs hatch, the female scuttles out to collect algae from the surrounding mud to feed to

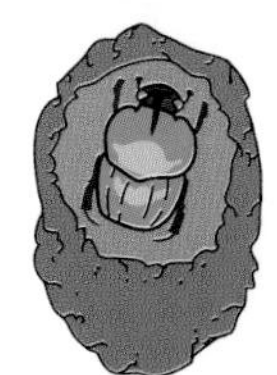

her brood until they're about a week old. The larvae then strike out on their own and excavate their own burrows.

Dung beetles have the thankless task of clearing away—you guessed it—dung! Dung is actually a very valuable commodity in the insect world as so many species feed on it or on the other species that do. The larvae of these beetles typically feed on nothing but dung, so the parents excavate elaborate tunnels and brood chambers under a mound of the stuff or nearby, and provision the brood chambers with lots of it. Male and females of these beetles are often equipped with brutal horns and other weapons to keep interloping rivals at bay. Depending on the species, the female beetle processes the dung often forming impressive orbs, the center of which has a cavity where the female will lay a single egg. When the orb is complete, she seals it with a plug of more fibrous dung and begins to work on more. When all of the brood balls are complete, the female stays with them, licking them from time to time, which is thought to impede the growth of fungi. She might remain in the brood chamber until the first of her young emerges as an adult.

∨ Many cockroaches are dedicated mothers. They shield, carry, and even suckle their young.

^ The breeding behavior of *Rhinastus* weevils has a dark side. The female weevil chews a hole into a bamboo internode and deposits a clutch of eggs. To begin with, her brood coexist quite nicely, eating the bamboo. Eventually though things turn nasty and one of the larvae starts cannibalizing its siblings until it is the only one left.

> Burying beetles are among the best-studied insect parents. They provision their young with a delicious corpse and even feed them.

Even though dung is decidedly moreish for these beetles, some of them have forsaken it. In some places in Amazonia in the late afternoon, enormous metallic blue beetles buzz past you at speed in a hurry to get somewhere. This is *Coprophanaeus lancifer*—a golf-ball-sized dung beetle that once seen is never forgotten. Instead of seeking out steaming piles of poo in the forest, it sniffs out corpses, stocking its subterranean brood chambers with hunks of rotting flesh that it chops from the carcass. I've seen the carcasses of mammals nearly stripped clean by these impressive beasts—all in the name of feeding their young. Other dung beetles have gone stranger still, provisioning their brood chambers with the new queens of leafcutter ants or even millipedes (*Canthon virens* and *Deltochilum valgum*). In both cases, the victims are clinically decapitated on the surface before being dragged underground.

∧ Some dung beetles provision their burrows with carrion rather than dung. In parts of the Amazon, *Coprophanaeus lancifer*, is among the first insects to arrive at carrion.

∧ Solitary wasps go to great lengths to build and stock a nest with paralyzed prey. Here, *Odynerus spinipes* is returning to its nest carrying a weevil larva.

Carrion, like dung, is a valuable commodity in the world of insects, and some insects that depend on it also put a lot of time and effort into the care of their young. Burying beetles depend on the corpses of small mammals and birds, which they locate using their acute sense of smell and promptly bury. The parents regurgitate partially digested bits of the carcass to give to their begging larvae. The parents tend them carefully until they're ready to pupate.

Some fungus beetles are on a par with the burying beetles in terms of parenting. The larvae of these beetles eat fungi, which are a sporadic resource. The female beetle stays with her larvae as they grow and actively shepherds them around to new areas of fungi.

The Lengths Some Mothers Go To

Among insects, wasps, ants, and bees show the greatest diversity of parental care. These insects have to seek out food that is patchily distributed, they must navigate to and from a nest that they stock with food and must interact with others of their kind—often huge numbers of them in elaborate nests. These challenges have driven the evolution of the complex behaviors we see in these animals.

Many of these insects—collectively known as the Hymenoptera—are solitary animals. To date, around 20,000 solitary hunting wasps have been described and they all use keen senses and venom to find and subdue their prey, which consists of other insects and spiders. Bees are essentially fluffy, vegetarian wasps. Instead of animal prey, they seek out nectar and pollen. Predator or not, these animals all provision a nest for their young that they make in all types of habitats, such as in sandy ground, in hollow plant stems, and in tunnels in dead wood. The amount of work that goes into making and stocking these nests is remarkable and, for this reason, the solitary wasps in particular are some of my favorite insects.

Very few of these have been studied in any detail and of these it is the beewolf that is probably the best understood.

EUROPEAN BEEWOLF

Philanthus triangulum

SOLITARY HUNTERS

- 135 species of beewolf
- All solitary hunting wasps go to great lengths to provision food for their young
- Males of nearly all the beewolf species establish and scent-mark small territories to try and win a mate

Sandy banks, small sand cliffs, and even paths on sandy ground are desirable real estate for the female beewolf who wastes no time after she emerges in the summer in digging a gently sloping tunnel, loosening the soil at the tunnel face with her jaws before kicking the spoil behind her like a cartoon dog after a bone. During these frantic excavations she might take a few breaks to drink nectar from surrounding flowers until she has a tunnel extending for about 7–12 inches (20–30 cm). This is no mean feat when you consider that she is only about 0.8 inches (2 cm) long. She then excavates the first of many brood chambers—anywhere between three and thirty-four. These small voids in which her offspring will develop—one larva to each brood chamber—are connected to the main tunnel.

The next step is where the beewolf lives up to its name for she must find her prey and the only food her larvae will ever know—the humble honey bee. Honey bees are powerful fliers, they're sharp-eyed and they sting, so catching and subduing them is far from easy. The female beewolf patrols flower-rich areas and with her own very sharp eyes she eventually spots her quarry. A honey bee sipping nectar from a flower is the perfect victim and this is when the beewolf strikes, grabbing the bee and bringing her sting to bear, piercing the thin membrane behind its front pair of legs to inject, with surgical precision, a tiny amount of venom into a cluster of nerve cells. The bee succumbs almost immediately, its muscles disabled by an irreversible paralysis. With the bee now helpless, the beewolf takes an opportunity to refresh itself in a slightly ghoulish manner by sucking the victim's mouth to drain the so-called honey-stomach of nectar. The beewolf clutches the now helpless bee to her underside and takes off for her nest, pushing her flight muscles to the limit and putting her navigation abilities to the test. The prey more than doubles the weight of the female and the flight back to the nest—perhaps more than a mile—is an extraordinary test of endurance.

Using the position of the sun in the sky and the location of landmarks, the beewolf makes it back to her nest and drags the bee into the darkness of the brood chamber. There, she licks the paralyzed honey bee all over, coating it with secretions from

^ A female beewolf (*Philanthus triangulum*) takes a paralyzed honey bee down into her subterranean nest.

special glands in her head. Initially, it was thought these secretions killed microbes, preventing the spoilage of the paralyzed honey bee before the beewolf larva got a chance to tuck in. It turns out the truth is much more interesting as these secretions effectively embalm the honey bee, filling in every little crack and smoothing out protuberances so there are no places for water to condense on. As there's nowhere for condensation to form the honey bee stays dry, and bacteria and fungi cannot thrive. The beewolf defeats fungi with physics.

The provisioning of the brood cell continues, each hard-won honey bee flown back to the nest from the sometimes distant

^ The beewolf (*Philanthus triangulum*) nest contains anywhere between three and 34 brood chambers, each of which is stocked with as many as six honey bees. In this brood chamber there are two honey bees and the sausage-shaped beewolf egg.

hunting grounds, until it contains as many as six honey bees. The mother beewolf can choose to lay a female egg or a male egg, and because the daughters are so much bigger than their brothers, their brood chambers will need to be stocked with more bees. When the brood chamber is fully stocked, the beewolf lays a single egg on one of the honey bees, seals the brood chamber and leaves her offspring to its own devices. After hatching from its egg, the larval beewolf makes short work of the paralyzed honey bees in its brood cell, and in as little as a week, the fat grub is ready to spin a silken cocoon in which it will see out the autumn and winter in a deep sleep.

2

TO EAT AND BE EATEN

Insects, like all other animals, must eat other organisms to survive, but the sheer diversity of their eating habits would make your head spin. There are few if any things in the terrestrial or freshwater domain that aren't on the menu for one insect or another. To do all of these insects justice would take a whole library of books, so what follows is a whistle-stop tour of the eating habits of insects.

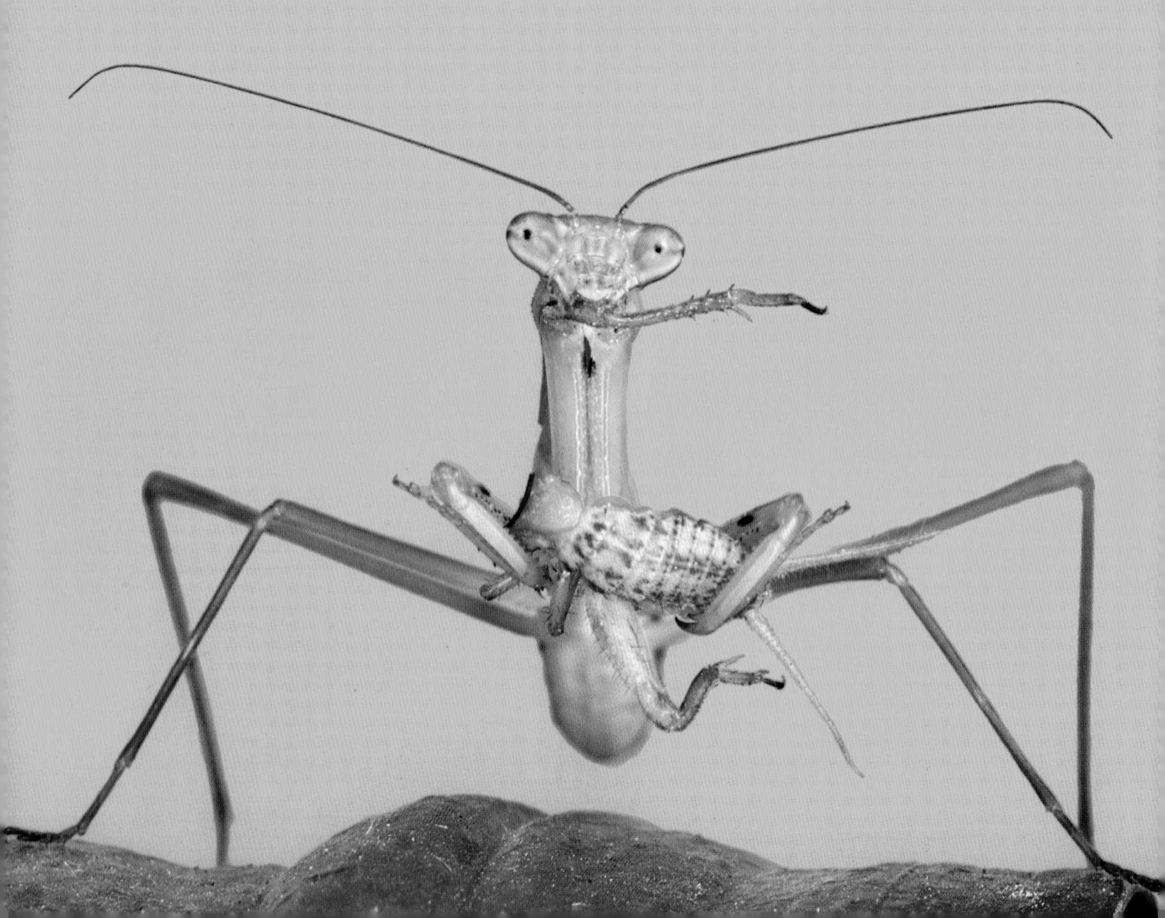

VEGETARIANS

Insects and plants go back a very long way—around 350 million years—and feeding on plants is by far the most common way of life among insects. The diversity of plants on Earth helps to explain the colossal number of insect species. In groups such as the grasshoppers and their relatives—true bugs, moths, and butterflies—the vast majority of species feed on plants.

Insect and flowering plant diversity exploded hand-in-hand at least 100 million years ago when plants worked out that insects make handy vehicles for the transportation of pollen. The beauty we see in flowering plants is not for us. All of the color and gaudiness is just advertising to entice the insects.

There are complex webs of insect life associated with all plants. Every green plant species will have at least one insect that eats nothing else and often there are several specialist insect species per plant species. No part of a plant is off limits either. Stems, leaves, flowers, seeds, roots, nectar, pollen, and sap are all greedily consumed by insects. Even aquatic plants do not escape the attention of insects.

< Preying mantises are extremely efficient ambush predators that use acute eyesight and raptorial forelegs to spot and catch a wide range of prey.

> Reed beetle larvae feed on the submerged parts of aquatic plants. They even tap into the plant's oxygen supply to breathe under water.

Aquatic Herbivores

The freshwater realm teems with herbivorous insects. The resplendent reed beetles feed on the roots of aquatic plants, such as water lilies, and can actually tap into the oxygen supply of the plant roots, so they can remain submerged for the entirety of their larval development. Many aquatic insects are specialist algal grazers, using modified mouthparts to graze from rocks often in raging torrents and clinging tenaciously to the slippery substrate using powerful limbs, claws, and even suction cups!

^ A huge variety of herbivorous insects use plant toxins to their own advantage, sequestering the chemicals to use them for defense against their own enemies. This is what the monarch butterfly does with milkweed toxins.

Plant Defenses

The chemical defenses produced by plants are enormously varied and many of them are used by humans—just think of caffeine, nicotine, and cocaine to name but a few. These chemicals can mess up insects in lots of ways. Some simply deter a hungry insect, whereas others interfere with the insect's development, making it lay fewer eggs or even killing it. Compounds such as tannins are very good at binding to proteins, so an insect feeding on leaves loaded with these chemicals may eat its fill, but may grow slowly or not at all as the tannins prevent the proteins from being digested properly. Remarkably, some plant chemicals are also distress beacons. When an insect starts feeding on a plant the chemicals that are released attract the natural enemies of the plant-feeder, such as parasitoid wasps. These come to the rescue and do terrible things to the oblivious insect as it carries on nibbling a leaf (see chapter 5).

Plant chemical defenses have been completely neutralized and even turned on their head by a huge variety of insect herbivores. Some plant-feeding insects use these chemicals to detect and find their food plants. The ultimate mockery of these defenses is by the insects that not only use these chemicals as a homing beacon, but are also able to assimilate them and even modify them for their own protection. The final insult is the broadcasting of the theft of these defenses in the bright warning colors of many insects. The warning colors are a conspicuous reminder of the struggle that rages between insects and plants.

Herbivore Niches

You could take a close look at any plant and find a whole community of insects centered around it. A small minority of insect species aren't fussy and will demolish every part of a plant, but others are more refined in their tastes, preferring to eat a specific part of their chosen food plants. The leaves might be nibbled by caterpillars, leaf beetles, and grasshoppers. Tunneling within the leaves will be various leaf-miners (see page 65), normally fly larvae or caterpillars, but also sawfly and beetle larvae.

The stem might be suckled by aphids and other bugs, as well as being mined by fly larvae, beetle larvae, and yet more caterpillars. The flowers will be nibbled by beetles and caterpillars, and seeds will be hollowed out or eaten in their entirety by specialist seed eaters. In the darkness of the soil, root specialists suck sap or nibble the roots.

Perhaps most bizarre of all are the diverse range of insects that hijack the genetic machinery of plants, forcing them to produce capsules of bounteous quantities of food in the form of galls (see pages 66–67).

Insects are small and their food plants might be very patchily distributed in the landscape—the odd plant or patch here and there—so they have to be adept at finding the plants they depend on. As any gardener will know, insects are incredibly good at finding their food plants. From a distance, the chemicals given off by a particular plant will attract insects, but as they draw closer, shapes and colors might also become important.

The heritage of the relationship between insects and plants is so old that there has been plenty of time for the emergence of evolutionary struggles with the insects and plants competing to outdo each other. A plant with nutritious leaves is so heavily targeted by herbivorous insects that it loses much of its capacity to photosynthesize. This, in turn, drives the evolution of various defenses, such as spikes, hairs, and toxins.

In response, insects evolve traits to neutralize the defenses, such as behaviors to strip off spikes and hairs, moving to less well protected parts of the plant or draining the toxins from a specific part of the plant.

> Sap-sucking insects, such as this lantern bug, can also load themselves up with plant toxins.

Microbe Partners

One of the downsides to plant feeding is that most plant tissues, with the exception of pollen and seeds, are very low in protein, so an insect must eat large amounts of this food and get it through its digestive tract quickly. Sap-suckers, such as aphids, have a real challenge on their hands because although sap is loaded with sugars it is almost devoid of protein and completely lacking in some amino acids that are essential for growth. To get around this problem, aphids have to swallow and process prodigious quantities of sap. Their digestive tract is geared up to remove the excess, sugary fluid and excrete it as droplets of honeydew. If you watch a colony of aphids you'll see tiny beads of honeydew forming on the back end of the aphids.

Drinking vast quantities of sap doesn't solve all their problems though, so these tiny sap-suckers have also enlisted the help of symbiotic microbes, which feed on the sap and produce the missing amino acids the aphids need.

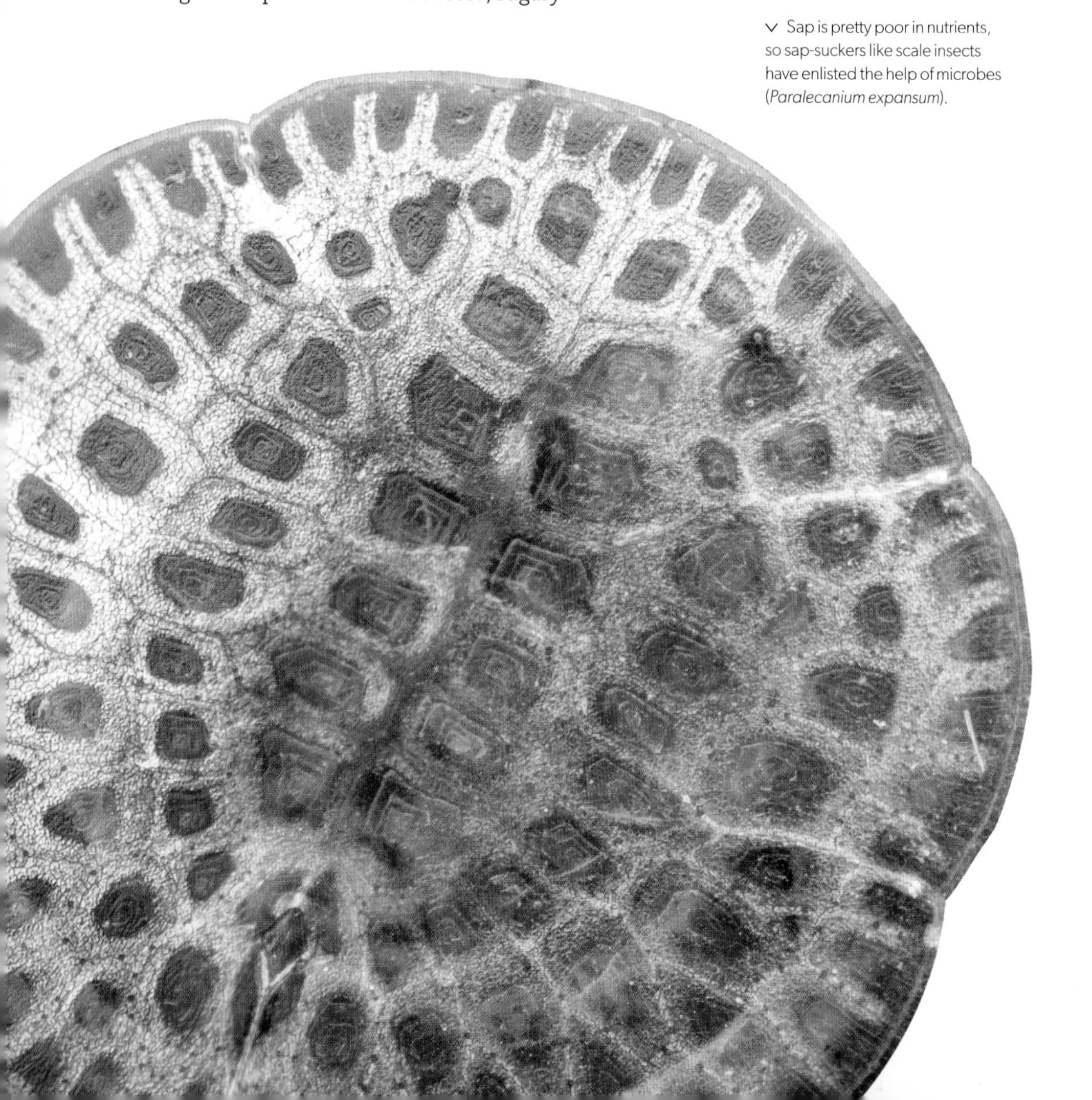

∨ Sap is pretty poor in nutrients, so sap-suckers like scale insects have enlisted the help of microbes (*Paralecanium expansum*).

LEAF-MINING INSECTS

You've probably noticed squiggles, scrawls, and blotches on leaves. These are the signatures of leaf miners, they are fascinating insects that feed within leaves. Thousands of species of moth, fly, sawfly, and beetle species live in this way. The moths and flies are the most diverse leaf miners and when you find some of the distinctive leaf marks the chances are that a fly or moth is the culprit.

∨ Leaf-mining is a popular niche in the insect world, especially among micro-moths.

In all these leaf-mining insects, the excavations are made by the larvae and in most cases they've become superbly adapted to a tunneling life style. They have reduced legs and eyes, and are often flattened to fit in the tight space between the outer layers of the leaf. Why do they do this though? It seems like a lot of effort compared with just sitting on the outside of a leaf and scoffing it. There are two main reasons. Firstly, insect larvae are plump and juicy, which many predators relish. Tunneling within the leaf, leaf miners get a degree of protection from these enemies. Secondly, the outer layers of the leaf are often brimming with chemicals that are intended to send leaf nibblers packing. By tunneling between the outer layers of the leaf they can access the plant's sap and less well defended cells. Many leaf-mining insect larvae even have wedge-shaped heads to separate the epidermal layers of the leaf.

Even these cunning tricks to avoid the plant's defenses are sometimes not enough though. Many plants have latex cells and if these are ruptured by the leaf miner it might be drowned in its tunnel. There are even specialist predators and parasitoids that have adapted to feed on leaf-mining larvae—either extracting them from their tunnels or, more fiendishly, injecting eggs into the hapless leaf miner through a needle-like ovipositor. There's also a waste problem with this way of life. All that succulent leaf tissue means a lot of droppings (frass) and the leaf miner either packs these densely into the tunnel behind itself or pushes them out of "mine shafts" that they chew to the outside world.

GALLS

At some point or another you've probably seen strange structures on a plant that look like they don't really belong there. These are galls and, as well as insects, they can be the work of viruses, bacteria, fungi, nematodes, and mites. They can range in complexity from simple curls or pouches to complex organs where one type of plant tissue has been forced to differentiate into many. The ability to make galls has evolved independently in lots of insects, but most of them are the work of midges or wasps. The way in which these insects force the plant to make these structures is another example of the ancient relationship between insects and plants and the beguiling complexity of nature. We've known about galls for thousands of years, yet exactly how they form is still a mystery.

What we do know is that in the case of insect galls, some stimulus from the insect makes an actively growing part of the plant, such as the bud, switch from its normal path of development to an abnormal path that ends in the formation of a nice, sheltered den made of food that nourishes the young of the gall-maker. Typically, a gall insect only goes for a specific part—often the developing leaves—of one species of plant and the structures they create are distinctive enough to allow the

^ A marble gall wasp larva inside its gall (*Andricus kollari*).

^ The elaborate Robin's pincushion gall is caused by the larvae of the gall wasp (*Dipoloepis rosae*).

identification of the culprit. The stimulus could be venom or other compounds injected by the female when she's laying her eggs or perhaps secretions from the newly hatched larvae. We just don't know. Some experiments have suggested that the stimulus from the insect is somehow turning on and harnessing some of the genetic machinery that is responsible for the formation and flowers and fruits.

INSECTS THAT CREATE GALLS

- Colonial aphids
- Thrips
- Gall wasps
- Gall midges

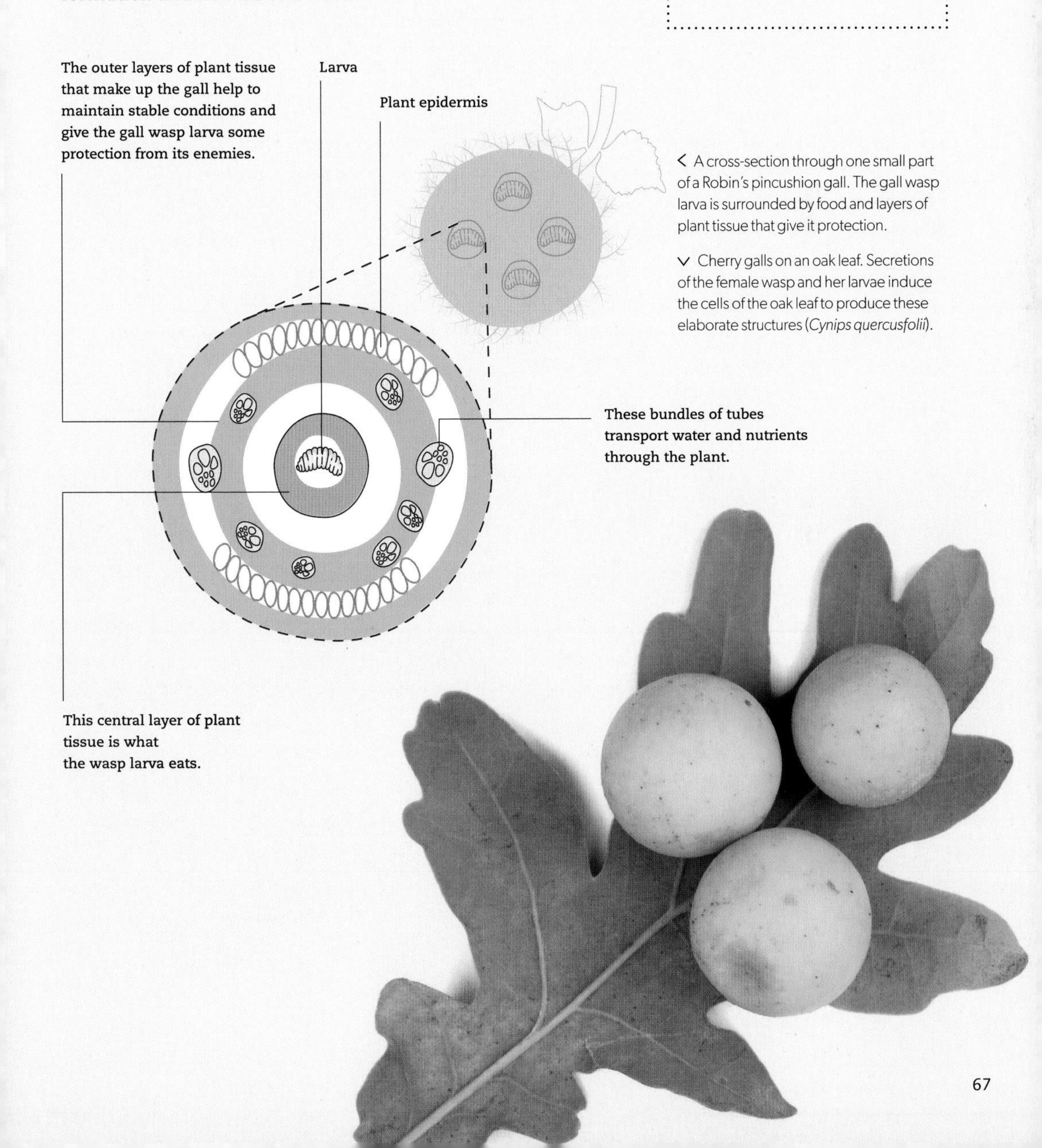

< A cross-section through one small part of a Robin's pincushion gall. The gall wasp larva is surrounded by food and layers of plant tissue that give it protection.

∨ Cherry galls on an oak leaf. Secretions of the female wasp and her larvae induce the cells of the oak leaf to produce these elaborate structures (*Cynips quercusfolii*).

POLLINATION AND SEED DISPERSAL

The relationship between plants and insects is complex and full of mystery. However, the aspect of this ancient association that is perhaps most familiar to us is pollination. We've all seen bees and many other insects with their head buried in a flower, it's such a common-place scene that few of us probably stop to think how special it is. Pollination is one of the most important ecological interactions on land.

In visiting plants for nectar and pollen, insects have become a crucial part in the reproduction of most plant species. When it comes to reproduction, plants are in a difficult situation because of their immobility. Without pollinators their options for sexual reproduction are limited. They can self-pollinate, but do this enough and the plants risk the loss of genetic diversity and the perils of inbreeding. Alternatively, they can rely on the wind and water to move the pollen to where it's needed. This can be a successful strategy—just look at the grasses and all the tree species that are wind pollinated; however, it can be a bit

∨ Flies are probably the most significant pollinating insects. Clinging to the mouthparts and furry thorax of this hoverfly are lots of pollen grains (*Syrphus* sp.).

> Plant engineering—when pollen is in short supply and the plants near their nest are not yet flowering, bumblebees will nibble holes in their leaves. The damage stimulates the plants to flower weeks ahead of schedule, providing the insects with welcome food when flowers might be in short supply.

haphazard with a lot of wastage. Animals of all sorts, mostly insects offer a much more precise service—taking pollen to exactly where it's needed, although enticements are required.

Nearly 90 percent of flowering plants are pollinated by animals and most of these are insects. The relationship between these plants and their pollinators has become so intertwined that on both sides remarkable adaptations have evolved to make the process ever more efficient. Just think of the energy and materials that a plant puts into its flowers and the rewards it must lay on for its pollinators. Some plants can even boost the memory of their pollinators by providing a small hit of caffeine in their nectar. Look at the adaptations of pollinators, such as their sharp senses for finding the flowers and their specially modified structures for accessing and carrying the rewards, such as the long, straw-like mouthparts of moths, flies, and bees, as well as the various types of pollen "baskets."

Pollination is synonymous with bees, and although these insects are very important pollinators they're not the only ones. In terms of the number of species that visit flowers, the moths and butterflies are number one (about 140,000 species), with the beetles, and wasps, ants, and bees equal second (about 77,000 species each), and the flies third (about 55,000 species). Insects in other groups also visit flowers, but these are all quite minor in terms of species compared with the above. It's likely that when we better understand what species are out there and which insects are pollinating which plants, the flies will be at number one with more pollinator species than any other group of insects.

Working out which of these groups of insects are the most effective pollinators is even more difficult, but the top spot here probably goes to the bees and second place to the flies if we're talking about what proportion of plants in a given community are visited by a particular group of insects. This is no surprise as bees are pollination specialists. At least 130 million years ago, a group of solitary hunting wasps started provisioning their nests with pollen—perhaps tiny, accidental amounts to begin with—but this is how bees started out, they're really just hairy, vegetarian wasps. Hairy to better trap pollen and vegetarian, because all of the 17,000 known bee species feed more-or-less exclusively on nectar and pollen.

As well as being very familiar we also attach a certain amount of simplicity to pollination, but as with every other interaction in nature the relationship between plants and their pollinators is dynamic and complex. For one thing, nectar isn't cheap to produce. Plants can't just slosh about gallons of water and sugars for free. These are rewards and only the most efficient pollinator should be able to access them. This is why flowers often have deep nectaries, accessible only with a sufficiently long straw, which is what the mouthparts of many specialist pollinators have become.

After pollination, an insect's job is done. However, there are a few plants, such as the *Corymbia torelliana*, that have also enlisted insects to disperse their seeds. The seeds of this tree contain a sticky resin that is coveted by certain stingless bees who use it for nest building. Bees collecting the resin also end up carrying off some of the sticky seeds. Perhaps the most elaborate ruse must go to the plant *Ceratocaryum argenteum* from South Africa. The seeds of this plant look and smell like pellets of dung and are enough to dupe dung beetles into rolling them away and burying them.

∨ The Yucca moth has to trade off between pollination and larval development. Female moths pollinate the flower, but then lays eggs into some of the developing plant seeds.

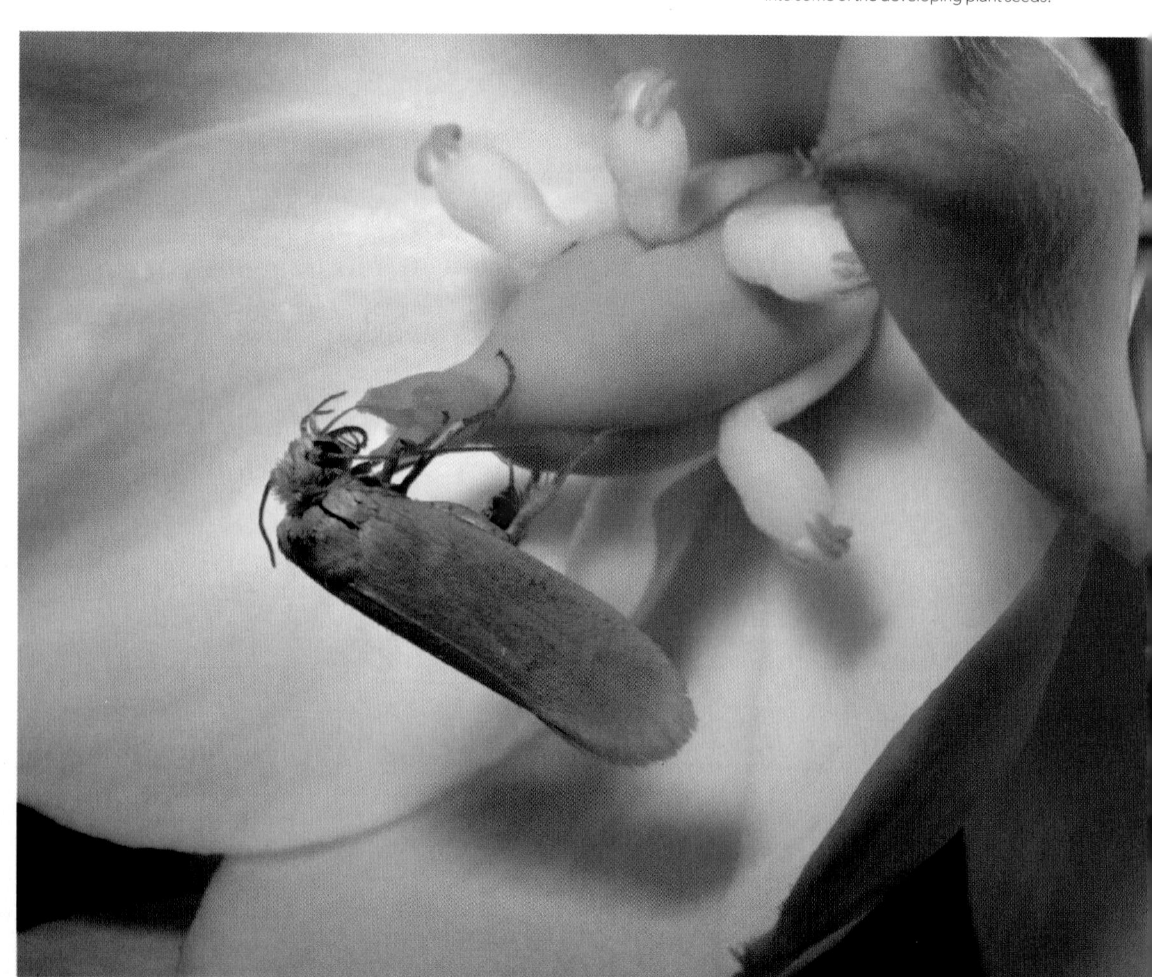

∧ Some dung beetles in South Africa have been duped into dispersing the seeds of the plant *Ceratocaryum argenteum*. These seeds smell like dung and the beetles roll them away and bury them.

∨ This image shows one of the duped dung beetles (*Epirinus flagellatus*), a *Ceratocaryum argenteum* seed and Bontebok dung—the type of dung they go for when not being fooled by the plant.

CARNIVORES

Most insect species might be plant feeders, but in the enormous diversity of these animals there's plenty of room for a mind-boggling diversity of predators. Bear in mind also that who eats who and exactly how they do it is full of unknowns, so just think of all the things that are still to be discovered.

For the most part, predatory insects feed on other insects, or other arthropods. There are some remarkable exceptions to this, which we'll go into later. The predators range from ambush specialists who wait for prey to come to them to very active hunters who scour their habitat for prey.

∨ The larvae of the click beetle, *Pyrophorus nyctophanus*, produce a green light to attract prey. Lots of them have made their burrows on this termite mound in Brazil.

^ The mouthparts of an antlion larva make a fearsome trap and snap shut on any prey that blunders too close.

Ambush

For an insect, the world is full of terrors with a bewildering variety of ambush predators lurking in all manner of places. The larvae of owlflies (*Ascalaphidae*) lie in wait on the ground or on trees, their enormous, hypodermic mouthparts agape and ready to snap shut when prey ventures too close. To enhance this strategy, their camouflage is superb—muted colors blending in with the background and a flattened body fringed with bristly outgrowths to break up the outline. Owlfly larvae subdue their prey with toxins, weirdly though, these appear to be produced in their gut rather than in specialized venom glands.

Relatives of owlflies—the antlions (*Myrmeleontidae*)—have gone one step further. Instead of just camouflage, some antlion larvae sit in wait at the bottom of a funnel-shaped pit in sandy ground to enhance their ambush. When a small insect like an ant blunders into the pit, it careers down the slope and into the waiting jaws of the antlion. Often, the prey makes a bid for freedom, but the task of scaling the loose sides of the pit is made more difficult by the antlion bombarding the prey with showers of sand, which it flicks with its head. Inexorably, the hapless prey slides toward the antlion. In a wonderful example of convergent evolution, some fly larvae (Vermileonidae) have evolved a similar strategy. These so-called worm-lions sit curled up at the base of their pit and also hurl sand at struggling prey.

^ Tiger beetle larvae are masters of ambush. Waiting at the mouth of their burrow with jaws agape they lurch explosively at any insect or spider that wanders too close.

As larvae, many beetles are masters of ambush. Tiger beetle larvae excavate vertical shafts in the ground or in dead wood in trees, plugging the entrance to their tunnel with a reinforced head. They wait motionless in this position—black, beady eyes above their enormous open jaws. When prey comes within range, they lunge backward with lightning speed from their burrow to grab it before retreating quickly underground with their quarry. The difference between larval and adult insects in terms of form and function is beautifully illustrated by tiger beetles. The fossorial larvae are masters of ambush, whereas the stunning adults are active hunters—fleet of foot and large-eyed, and capable of incredible turns of speed.

To ambush prey in darkness, insects have employed light. The well-known Waitomo Caves in New Zealand are studded with eerie lights, shining from thousands of fungus gnat larvae (*Arachnocampa luminosa*). The lights lure other insects to a trap of sticky silk threads that dangle from the silk nests of the larvae on the cave ceilings. Less well-known are the termite mounds of South America similarly studded with light. In this case, each light is the work of a click beetle larva that shines to lure prey. Some fireflies have moved away from just using light for the purposes of courtship (see previous chapter). These femme fatale species mimic the light display of other female fireflies, luring the males of these species to their doom as they drop in to investigate what they think is a receptive female.

Luring prey within reach of a lightning-fast lunge is also used by insects that can't put on a light show. Not all that many examples of this are known, but the more we understand about the different odors produced by insects, the more examples will come to light. The predatory rove beetle, *Leistotrophus versicolor* (see page 40), secretes a stinky liquid from its rear-end, daubing it on leaves to draw in filth-loving flies.

The resin assassin bug (*Amulius malayus*) also sets a trap for its prey, by dipping its forelegs in sticky tree resin. Stingless bees can't get enough of this stuff, but trying to get some from the assassin bug's legs will not end well. The resin gives added grip to the bug's lunge and the bee ends up impaled on its mouthparts.

Inhabiting the same southeast Asian forests as the resin assassin bug is the fabulous orchid mantis. It's not called this for nothing. You

∨ Owlflies are closely related to antlions and their larvae are also masters of ambush. Typically, they're beautifully camouflaged, sitting stock still on tree trunks and amongst leaf litter, waiting for a chance to strike.

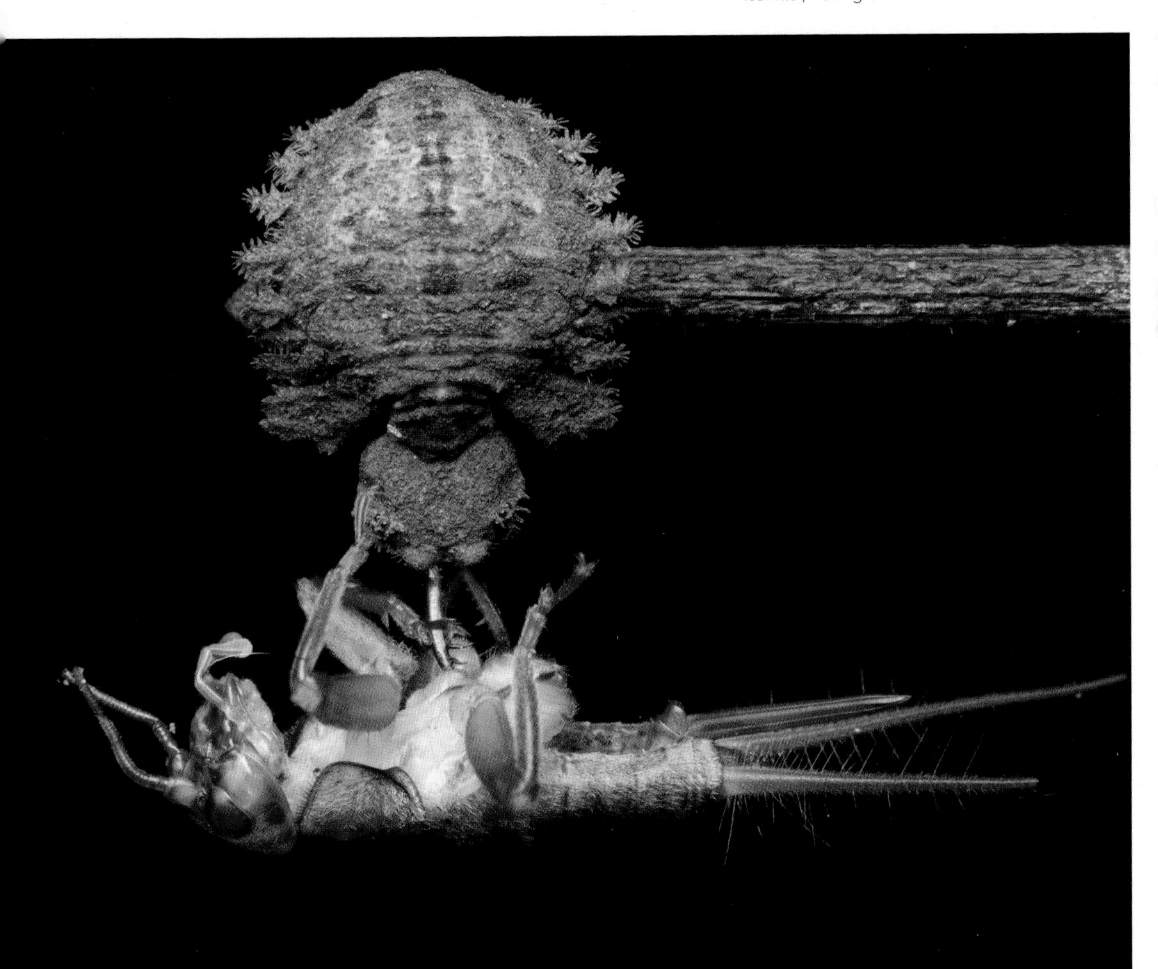

would be hard pressed to see one when it's sitting motionless on a cascade of orchid flowers. As well as superb camouflage against its enemies, the glorious appearance of this animal has a darker side too, because its floral mimicry actually attracts pollinator insects who think they are just visiting a flower for some nectar, but instead end up as the mantis's prey.

Perhaps the most remarkable example of prey-luring among the insects and one of the most fascinating examples of predation can be found in the *Epomis* beetles (see page 77).

∨ The beautiful orchid mantis mimics flowers in order to attract pollinators to their doom (*Hymenopus coronatus*).

EPOMIS BEETLES

The diminutive larvae of the *Epomis* beetles tackle prey way larger than themselves. Not only that, they choose to pick on vertebrates—frogs and toads to be exact. This is very unusual. Due to the obvious size differences there are few examples of insects predating vertebrates. This is my favorite and is testament to the dedication and observational skills of the biologists who wanted to learn more about the lives of these beetles. *Epomis* larvae lure a suitable victim by moving their mouthparts and antennae. The curious amphibian comes to investigate and eventually makes a lunge for what it thinks is an easy snack. It initially swallows the beetle larva, but its fate is now sealed as the larva has latched onto it with grappling hook mandibles and begins to feed, sometimes from inside the frog's stomach. Even adult *Epomis* beetles are partial to amphibians, latching onto a frog and consuming it.

∧ *Epomis* larvae have powerful, sickle-shaped mandibles for keeping a tenacious grip on their prey.

∨ An *Epomis* larva firmly attached to the throat of a toad. The larva will soon start to consume the unfortunate amphibian.

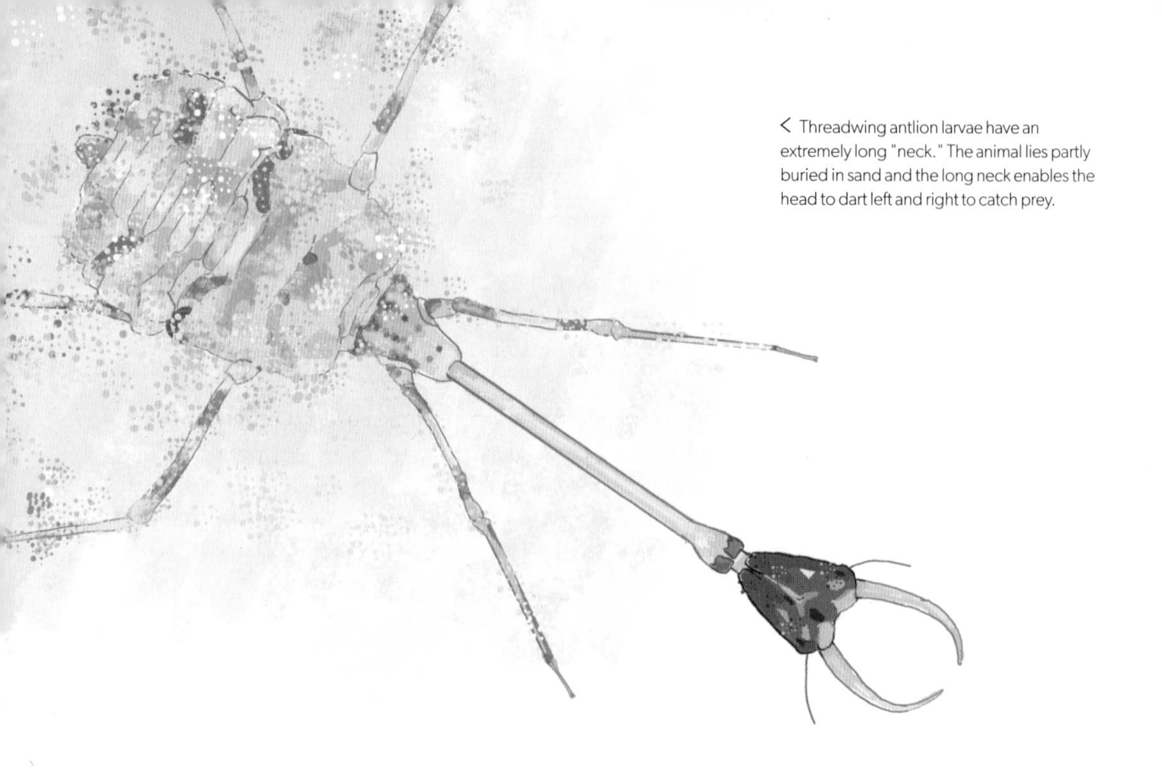

< Threadwing antlion larvae have an extremely long "neck." The animal lies partly buried in sand and the long neck enables the head to dart left and right to catch prey.

Ambush Adaptations

Ambush predation in all its many forms is often aided by adaptations that allow for very rapid snatching of prey. Just think of the raptorial forelegs of praying mantises and the raptorial forelegs that have convergently evolved in other groups of insects, such as the mantisflies, which are more closely related to lacewings and antlions. Raptorial limbs have also evolved in some flies—sometimes the first pair of legs, but also the second pair, which makes for quite an unusual death-grip. Lots of water bugs have extremely powerful raptorial forelimbs for bringing prey within reach of their pointy mouthparts—as I've found out when trying to handle some of the biggest species in the wild.

In addition to the special limbs for enhancing the art of ambush, lots of insects have evolved some impressive mouthparts that allow lightning fast strikes while the rest of the animal remains motionless. The nymphs of dragonflies and damselflies have hinged mouthparts tipped with grasping claws for snatching prey unawares. The nymphs wait motionless among aquatic vegetation for suitable prey to come within range of their secret weapon.

The larvae of threadwing antlions (*Nemopteridae*) have stuck their neck out, literally, in pursuit of perfecting the art of ambush. These rarely seen animals have evolved an extremely long neck for snagging prey while their plump body remains hidden in the fine sand of their desert home. This adaptation may also keep struggling prey away from their delicate body and allow the larvae to more safely find good sites to ambush from without falling prey to others of their kind.

Caterpillars are not really known for their meat-eating antics—the vast majority of them munch on plants quite happily. However, exceptions are the rule in nature and in splendid isolation on the islands of Hawaii, some caterpillars have turned ambush predators. Inchworms are good at mimicking twigs and carnivorous inchworms in the genus *Eupithecia* use this disguise to good effect. They wait motionless on plants and any unwary insect coming up behind them will blunder into a pair of thin appendages on the caterpillar's back end. This jolts the caterpillar into action, and it bends over backward to help the victim shuffle off this mortal coil. These caterpillars have well developed legs for grasping their prey and their mandibles are pre-adapted to deal with tough food stuffs as they evolved to chomp through tough leaves.

Across the Pacific in Panama, we find other ghoulish case-bearing caterpillars whose retreats are fashioned nearly entirely from the remains of their victims bound together with silk. *Perisceptis carnivora* attaches one end of its case to a surface, leaving the open end free from where it lunges out at passing insects.

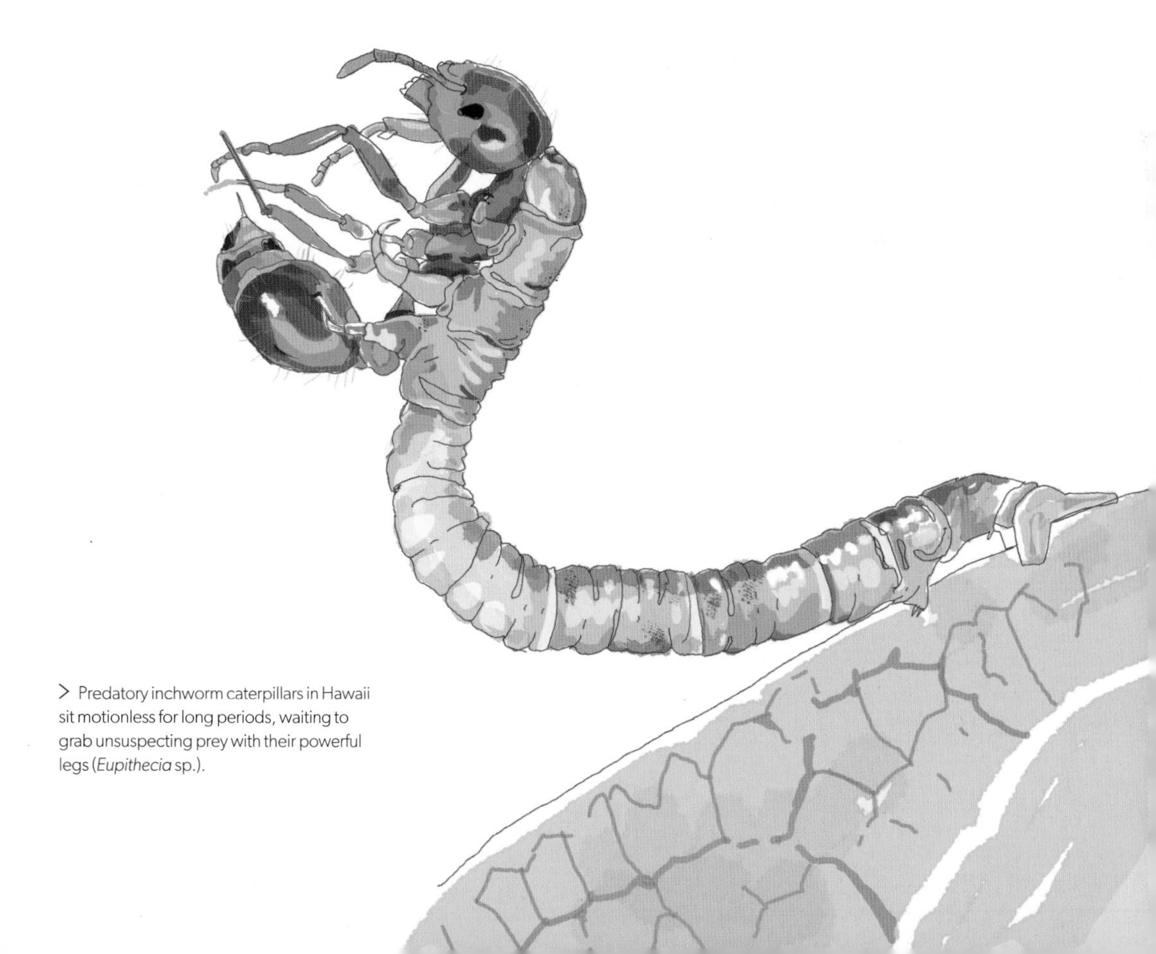

> Predatory inchworm caterpillars in Hawaii sit motionless for long periods, waiting to grab unsuspecting prey with their powerful legs (*Eupithecia* sp.).

THE HUNT

Ambush is an effective, low-energy way to catch prey, but a huge variety of insects have opted for a more active approach to predation. These prodigious hunters are everywhere, and even in your backyard there will be a dizzying array of them, from beetles with telescopic mouthparts to venomous flies.

The dragonflies and damselflies are the most ancient predatory insects; aerial assassins that have been plying their trade for at least 250 million years. These animals, especially the dragonflies are incredibly strong fliers and are furnished with the largest compound eyes of any insect, each of which is composed of around 28,000 separate units. Indeed, most of their head is eye and around 80 percent of their brain is devoted to making sense of the light information that streams in through the enormous peepers. Sharp senses and strong flight make dragonflies consummate predators. The hawkers patrol on the wing, scanning their surroundings for prey, while the darters watch for prey from a perch—darting

˅ Patrolling the air looking for prey requires acute vision and dragonflies have some of the largest compound eyes of any insect.

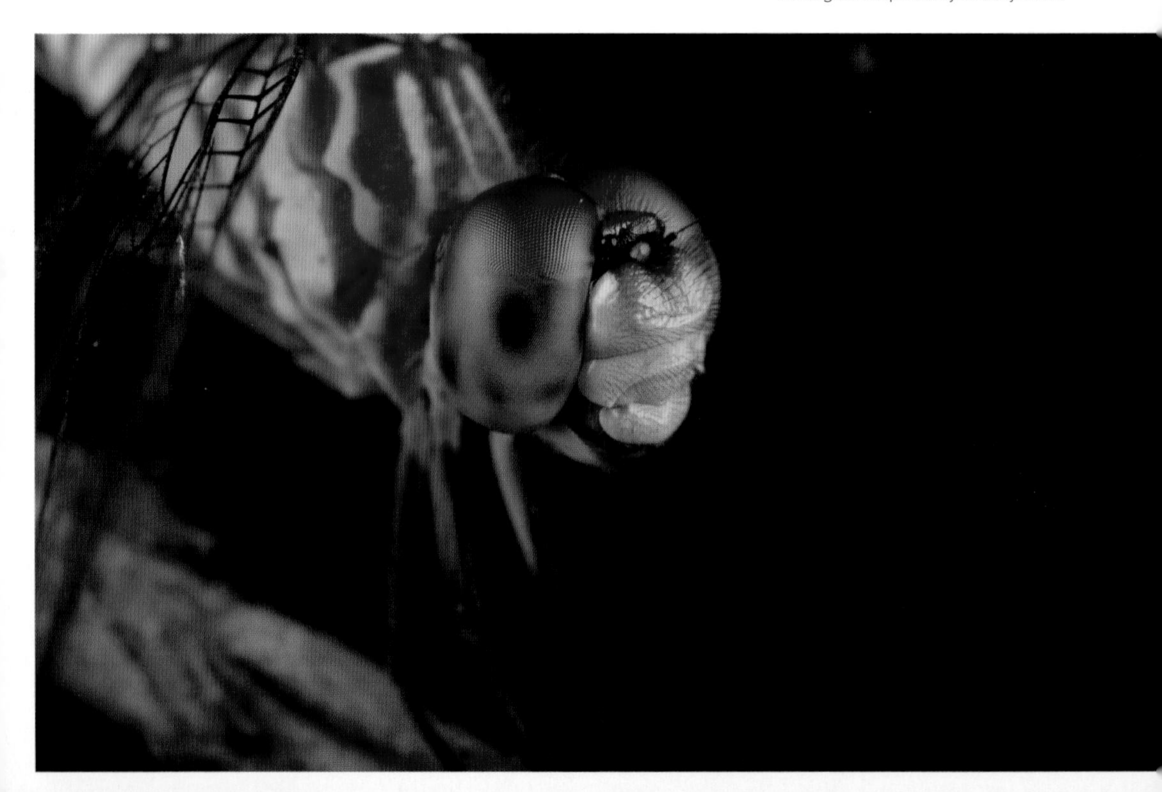

^ Assassin bugs impale prey and suck it dry using their straw-like, piercing mouthparts, known as a rostrum.

out to intercept it. As for their prey, they're not fussy, although small insects, especially flies, make up the bulk of their food. In all cases, the prey is snared with the spiny legs and then dispatched with the powerful mandibles.

Other notable winged hunters among the insects are the robber flies, also known as assassin flies. These range in size from a few millimeters long up to 3-inch (7-cm) monsters that rank among the largest flies. Like the dragonflies, they're very strong fliers and have enormous eyes for navigating at speed and detecting prey. Some robber flies are graceful hunters of grasslands, where they fly delicately among the tangle of vegetation, plucking unwary prey from the plant stems. Others are more brutish, taking their prey on the wing. Their piercing mouthparts inject potent venom into the prey, which quickly kills it and turns its insides into a smoothie the fly can suck up (see page 83). With their sharp mouthparts and fast-acting venom, robber flies can tackle heavily armored prey, such as dung beetles. They're also equipped with bristles, including a dense mustache on their face that protects them from being damaged by struggling prey. The larvae of these flies are also venomous predators, but we know very little about them. Generally, the larvae are animals of the soil or dead wood, where they hunt a range of invertebrates.

As already mentioned, adult tiger beetles are formidable hunters. With enormous eyes, brutal mandibles, and an incredible turn of speed—both on the ground and in the air—they're able to take on all sorts of prey. They're so fast—the fastest cover 120 body lengths in a second—that their vision simply can't keep up when they're running at full tilt; their surroundings become a featureless blur. To get around this problem they have to stop periodically, pinpoint their quarry again, and then continue the chase. Not only that, but their antennae allow them to detect obstacles, giving their brains the information it needs to pull evasive maneuvers or simply skitter across the top of small obstacles. Tiger beetles are found around the world and some species are fairly easy to spot as most of them are active during the day and their stop-start hunting technique is a real giveaway.

Tiger beetles are pretty special, but some of the most niche hunters in the insect world have evolved to exploit prey that other animals leave well alone. Millipedes are among the best protected of all terrestrial animals. Not only do they have a tough exoskeleton, but many of them produce copious amounts of weird and wonderful toxins, such as 1,4-benzoquinones, phenols, hydrogen cyanide, quinazolinones, and alkaloids. All of this is not enough though. Among the beetles there are a few species that are specialist predators of millipedes. The larvae of railroad beetles are glowing (they emit light too) cylinders of malice that seek out and consume millipedes. The larvae of some railroad beetles can be up to 2.5 inches (6.5 cm)

∨ Railroad beetle larvae are millipede specialists, hunting them down and dispatching them with clinical efficiency.

long and they trundle frantically across the ground in search of their prey. When they find a millipede they tackle it and pierce its tough exoskeleton using their hollow, sickle-shaped mandibles. At this point, it's all over for the millipede as the beetle larva proceeds to pump gastric juices into the prey, which somehow disable the millipede's ability to discharge its chemical defenses from the glands along its body. Following this initial assault, the beetle larva retires to a safe distance, buries itself in the ground and waits a good while. The fluids it injected liquify the millipede's insides and the predator eventually returns to slurp up the contents of each of the victim's segments, starting at the head end.

^ The stings of wasps, ants, and bees contain a cocktail of compounds that are used in predation and defense (*Philanthus triangulum*).

Venom and Prey

A huge variety of insects use venom to subdue their prey. Indeed, venom is such a versatile tool that it has evolved independently at least fourteen times in insects, including true bugs, aphids, lacewings, beetles, wasps, bees, ants, some caterpillars, and flies. In most of these groups, the venom is produced by glands associated with the mouthparts. The notable exception here is the venom dispensed by wasps, bees, and ants, which is produced by sex glands and injected using a sting derived from an egg-laying tube (ovipositor).

Venom is not one thing. Rather, in any given species of venomous animal it is a cocktail of compounds—often hundreds—all of which have a particular role. Some might just kill the prey; some might start breaking the prey down from inside—giving digestion a head start or even completing this process—and turning the prey into a smoothie in those venomous insects that have sucking mouthparts. Other compounds in the venom quickly paralyze the prey, because the last thing any self-respecting predator wants is a victim thrashing around for ages. These paralyzing compounds are very important in all sorts of wasps that need fresh, inactive prey for their offspring. The venoms of these wasps also contain compounds that modulate the metabolism and immune system of the prey.

STENUS ROVE BEETLES

HYPERDIVERSE ROVE BEETLES

Stenus is one of the largest genera of animals:
- 3,000 species worldwide

Stenus rove beetles are charismatic little insects. Very common in the right habitats (among moss or tussocky vegetation), they are easily overlooked. These beetles are large-eyed, active predators of even smaller arthropods, such as mites and springtails. Their secret weapon for catching such quarry is telescopic mouthparts. A section of their mouthparts—the labium—can be shot out extremely rapidly by forcing hemolymph into it. In absolute terms, the distance it can squeeze this structure out to is nothing to write home about, but it can extend to at least half the beetle's body length, and this is the difference between securing a meal or going hungry.

The business-end of this telescopic labium is studded with hair-like setae and pores that secrete an adhesive substance to ensure the prey is well and truly snared before it's yanked back to the sharp mandibles of the beetle.

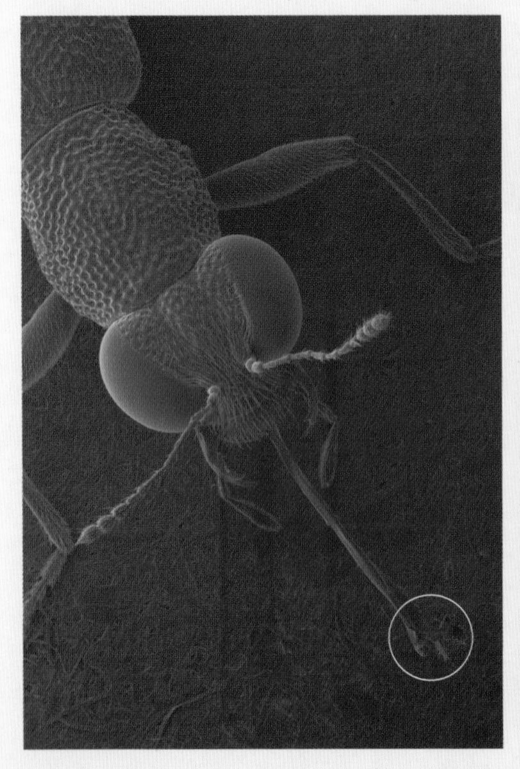

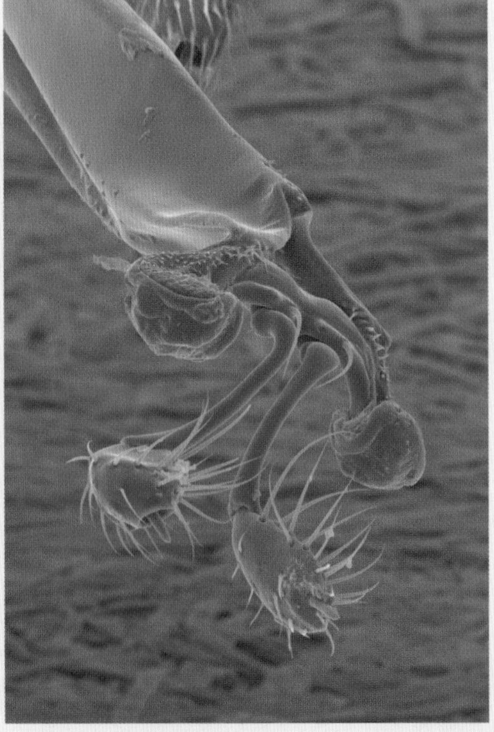

∨ *Stenus* rove beetles have telescopic mouthparts that can be shot out under fluid pressure to snag prey. The tip of this weapon is bristly and secretes adhesive substances.

GRUBBY HUNTERS

The larvae of many types of fly are active predators. Many of them spend most of their lives as ferocious hunters of aphids, moving with surprising speed on leaves and stems searching for their juicy prey. Likewise, horsefly larvae are also active predators on boggy soil or in freshwater. For the most part though, we know next to nothing about the lives of young flies. They often live out of sight in micro-habitats that are very difficult to study without destroying them. It is likely that a good proportion of the fly species, where the larval biology is unknown, are active predators. Get out there and make your contribution to entomology by finding out where and how these animals live!

∨ The fearsome mouth-hooks of a horsefly larva. These are active, venomous predators in freshwater and very wet ground. They are even able to tackle small vertebrates.

HUNTING IN FRESHWATER

Some predatory insects have also carved out unusual niches in freshwater. The so-called spongeflies (they're neither sponge nor fly) are relatives of lacewings and antlions, and they've evolved to exploit the most unusual prey—freshwater sponges and another group of strange, sessile aquatic animals known as bryzoans. Sponges are strange ancient animals, often loaded with noxious chemicals and unpalatable spicules. Likewise, bryozoans are far from being a delicacy. Still, spongefly larvae can't get enough of them and, on the plus side, sponges and bryozoans can't exactly run away. The spongefly larvae probe their prey with long, flexible mouthparts—sucking up the contents of cells with impunity. This is quite some niche, which says a lot about the dazzling diversity of insect lives. You would even be forgiven for thinking that living in this way would free you from the clutches of the many predators of the terrestrial domain. Unfortunately, on their path to an aquatic existence, the spongeflies were followed. Like every insect species they have parasitoids—other insects that seek them out like bloodhounds and make short work of them. More of which—much more—later. . .

∨ Spongefly larvae (L) are specialist predators of freshwater sponges. They use their long flexible mouthparts to drain the contents of sponge cells. The adults (R) are relatively short-lived terrestrial animals.

Snail Specialists

Snails are generally well-defended animals, but there are plenty of insects that have become snail connoisseurs. Rather than the uncouth and primitive snail-eating habits of some birds who simply smash the snail's shell into pieces, insect snail hunters dispatch their prey in precise, albeit appalling ways.

Natural selection on the Hawaiian Islands has also produced predatory, case-bearing caterpillars that seek out and eat tiny land snails. When one of these snail hunters finds a likely victim, it binds it in place with silken threads from the silk glands in its head. The caterpillar then wedges the edge of its case under the snail's shell and pursues the mollusk as it retreats deeper into what was once its fortress. After the snail has been devoured, the caterpillar may even incorporate the empty shell into its own case.

Many types of beetle feed exclusively on snails, both as larva and adult. Their heads and the front parts of their bodies are often slim, allowing them to reach deep into the shell to feed on the quivering mollusk. Another favored tactic of these snail-eaters is the regurgitation of lashings of gastric juices into the snail's shell, turning the occupant into a nutritious gloop in its own handy container, which can be greedily slurped by the predator. There are even some tiny dung beetles that depend on snails, but they feed only on snail slime. What the larvae of these dung beetles get up to is unknown.

Killer Farts

Before we leave the predators, there is one final species I need to tell you about. The larvae of a type of beaded lacewing (*Lomamyia latipennis*) nobble their victims with toxic "farts." These flatulent characters live in termite nests, preying on the termites, which they subdue by releasing toxic vapors from their anus. One puff of this stuff is enough to immobilize six termites in one go, which the lacewing larva promptly tucks into. You can guarantee that there are countless more, equally bizarre morsels of insect biology out there just waiting to be discovered.

< At least one species of beaded lacewing emits toxic gases from its anus to subdue its prey—termites (*Lomamyia latipennis*).

THE CLEAN-UP CREW

Earth is bristling with life and all that life means lots of waste. Just think of all the living things that are continually shedding bits all over the place, excreting, defecating, and dying. All of this life means colossal amounts of waste. Waste that needs to be recycled and first in the queue to get stuck in are the insects.

A large animal dies and no sooner has it breathed its last than insects arrive in their droves, clamoring to claim their piece of the corpse. A tree falls over in the forest and an army of wood and fungi munchers will converge to return the giant to the soil. An elephant drops a serious mound of poo and insects, great and small, come flocking, eager to bury themselves in it and eat their fill.

Welcome, then, to the glamorous world of the scavengers. They'll make you wretch, they'll make you shudder, they'll put you right off your dinner, but without them we'd be in a right old fix knee deep in all manner of filth. All of the many insects with a proclivity for dirt and decay are unloved and generally overlooked, but their task is crucial.

The extent to which insects have diversified to make use of what we see as downright unsavory habitats is perhaps best illustrated by some of the flies. Take the bone-skipper fly (*Thyreophora cynophile*)—well, if you can find one. This almost mythical fly deserves to be celebrated for many reasons. Not only does it look pretty funky, but it was thought to be extinct for about 160 years until one turned up in Spain in 2009. It also develops exclusively in the marrow cavities of the bones of large mammals. Again, like other insects we have seen so far, this is quite some niche, which underlines how far the insects will go in the name of cleanliness.

Corpses of large animals are attended on their journey back to the soil by a fleet of insects, all of which are specialists at dealing with specific types of waste. Flies of various types, such as blow flies, deal with the corpse in the early stages of decay. The larvae of these flies—maggots—are corpse-strippers *par excellence*, devouring all the soft tissues in a remarkably short period of time, especially in warm weather. When all the soft material has been consumed, other insects, including various beetles and even moth caterpillars, take up the mantle, gnawing on the skin and sinewy bits until just bones, scraps of skin, and tassels of sinew remain. The efficiency with which these insects reduce an enormous animal, such as a cow or an elephant, to bare bones is something at which to marvel. When small mammals and birds reach the end of the line, burying beetles, which we've already been introduced to because of their parenting abilities, see to them in their entirety, interring them out of the way of other corpse connoisseurs.

< A huge variety of fly larvae consume dead and decaying organic matter making them recyclers extraordinaire.

∨ Thought to be extinct for about 160 years, the bone-skipper fly depends on the carcasses of large mammals as the larvae develop in the marrow cavities (*Thyreophora cynophile*).

In addition to all the corpses, Earth also has a bit of a poo problem. There are a lot of animals producing prodigious quantities of solid waste. In 2014, it was estimated that humans and livestock produce just over 4.4 billion tons (4 billion tonnes) of feces, this is more than ten times the combined mass of all humans. Worryingly, this is expected to rise over 5.5 billion tons (5 billion tonnes) by 2030. Luckily, the insects are on hand to gobble it up. This material doesn't seem very promising, after all, it's waste. However, it's still loaded with nutrients, which is why so many insects and other organisms love it so much.

Dung beetles are masters at dealing with this stuff (see pages 92–93), but lots of other insects go crazy for it too, loitering near where it emerges to claim it first. Most anxious of all is a certain beetle (*Canthon quadriguttatus*) from Amazonia. This little dung beetle spends a rather disturbing amount of time clinging to the fur around a monkey's bum in the hope of getting first choice of the steaming simian turds. When the monkey eventually gives up the goods, the beetles hop on and ride them to the ground.

Dead and decaying wood represents another bounty for the recyclers. If you've ever been lucky enough to see a forest more-or-less unfettered by human hands, you'll see dead wood everywhere. Trees die and remain standing, some fall, limbs and branches die in the canopy or crash to the ground in storms. All of these, in every situation, are valuable real estate to a whole, complex web of life. The type of wood, its diameter, moisture content, position, and fungal diversity all have a bearing on which animals

∨ A huge variety of insects depend on the dung of larger animals for food. These recyclers are crucial in terrestrial and freshwater ecosystems.

^ As their name suggests, giant timber flies are among the largest flies. The larvae develop in dead and dying trees.

will be using a particular piece of dead wood. Among the insects, the beetles and flies are masters of this domain, although some of the wasps are specialist wood munchers. Many of the insects that depend on dead wood are actually there for the fungi. Fungi are the engines that break up wood, freeing locked-up nutrients and energy, and returning them to the soil to fuel more plant growth.

The most important recyclers among insects are flies. There's no muck too disgusting, no filth too rancid, no offal to awful for these animals. A huge variety of flies do all their larval development in rotting organic matter and lap at it as adults. Still more flies are predators, drawn to waste to feed on the gathered insect masses. The largest flies of all—the timber flies—complete their larval development in dead and moribund trees, eating a mixture of wood and the microbes that feed on the wood. A tree that offers just the right conditions for these larvae may be home to many hundreds of them; the sound of their collective munching can be audible from a few feet away. Oozing from the wounds of injured trees is often a mixture of pungent sap and fungi, known as slime-flux, which has its own gang of fly specialists. There are even flies that are solely dependent on millipede dung for sustenance as larvae. Females of these flies cling to the millipede, eagerly awaiting a deposit. The largest compost heaps on Earth are the strands of seaweed that are deposited by the tides on every coast. Flies—most of which are found only in seaweed—and microorganisms are responsible for recycling this material, they break it down into nutrients that are reclaimed by the ocean at the next high tide.

DUNG BEETLES

We've already seen that dung beetles go to some length to ensure their offspring are well provisioned, which makes them crucially important in terrestrial ecosystems. They take dung straight into the soil from where it's deposited, or ball it up and roll it off for subsequent burying. Dung beetles are key players in sorting out this waste because they love to dig. They break up the dung, redistribute it, and drag it down into the soil. When their larvae have gleaned what they can from the dung, their own poo and what remains of the dung enriches the soil, and their digging also aerates and improves the drainage in the soil.

Dragging all this organic material into the soil and churning it enhances plant growth as much as or even more so than chemical fertilizers! There's more to dung beetles than just nutrient cycling though. They also disperse seeds within the dung and bury them, and because they eat this waste, they also demolish many of the parasites it contains and make it unavailable to other animals that covet it, such as flies. This helps to keep the populations of parasites and flies in check.

∨ Not all dung beetles roll balls of the stuff away to bury. Lots breed within dung and many more excavate their brood chambers directly underneath it.

The activities of dung beetles, or rather the lack of them, were felt acutely in Australia in the first half of the twentieth century. The native dung beetles in Australia had evolved to deal with the waste of marsupials and the copious feces of introduced cattle was not to their liking. The cow pats sat around on the surface and the populations of flies, many of which spread diseases, reached plague proportions, hence the now famous cork hats of the Australian Outback. Luckily, other dung beetles were coming to the rescue. Between 1967 and 1982, fifty-five species of dung beetles, many from South Africa, were imported and released with spectacular results—cork hats are just for tourists these days.

Some dung beetles have forsaken this resource. A few of them are important pollinators of plants that have an aroma of decay and the populations of some leafcutter ants—among the most important herbivores in the New World—are regulated by predatory dung beetles that feed on the new queens as they're trying to establish a nest.

A LIFE WITH DUNG

- Some dung beetle species have been shown to navigate and orient themselves using the Milky Way
- Depending on the species, these beetles roll the dung away, tunnel underneath it or live within it
- 9,500 species of dung beetle

∨ For a dung beetle, *Canthon virens* has gone on something of a tangent. Rather than dung it seeks out new queen leafcutter ants, decapitates them, and squirrels the body away underground as food for its offspring.

3

DEFENSES

To say that insects have lots of enemies would be an understatement. Birds demolish them by the beak load, mammals and amphibians gobble them up with gusto, spiders and other arachnids think of little else, and that's before we even consider the huge variety of insects that eat other insects. To cope with this onslaught, it's no wonder that insects have evolved a bewildering variety of defenses: mimicry, armor-plating, explosions, metal-reinforced jaws, spines, and stings.

HIDING AND DISGUISES

An insect's first line of defense is simply to avoid detection and they excel at this. Many can play dead or live in places that are very difficult to get at, whether it's deep in the soil or in the heartwood of an ailing tree. True, specialist predators have evolved to hunt even the most well-hidden insects, but these hiding places offer a degree of protection from generalists.

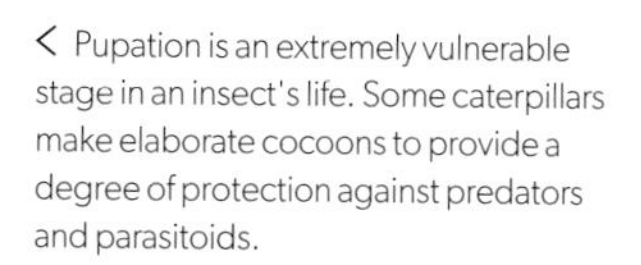

< Pupation is an extremely vulnerable stage in an insect's life. Some caterpillars make elaborate cocoons to provide a degree of protection against predators and parasitoids.

> The "fluff" and scales of moths can reduce the effectiveness of bat echolocation, making it harder for these predators to pinpoint their prey.

The insects known as spittlebugs have a cunning way of hiding. The nymphs of these insects slurp a serious amount of plant sap, but as they're small, soft, and stuck to the spot, they're easy pickings. To keep themselves hidden, they inject air into their excretions to make a thick coat of sticky bubbles that looks like spit, hence their common name. This dense coat of froth keeps them hidden from sight and conceals their smell to provide protection from some of their enemies.

Hiding can mean many things, not just keeping out of sight. Moths and bats have been locked in an ancient struggle for at least 50 million years, a struggle that has generated all sorts of remarkable adaptations. Since bats "see" with sound, giving them the slip takes on a whole new meaning. Some moths keep out of sight by flying very close to vegetation, so bats can't pick them out from the echoes of the background. The fluffiness of moths may also protect them from bats as it was recently discovered that the scales covering the body of some moths absorb up to 85 percent of the sound energy emitted by a hunting bat. This "stealth coating" can reduce the distance the bat is able to detect the moth from by around a quarter, which is the difference between life and death for the fluffy insect.

The long tails on the hind wings of some moths are more than just fancy ornamentation. These have been shown to fool bats into striking the wrong place. Indeed, bats attacking the beautiful long-tailed moon moths get a mouthful of tail most of the time. Perhaps the most extraordinary anti-bat adaptation is the sonar-jamming abilities of a tiger moth (*Bertholdia trigona*). When the moth hears the distinctive clicks of a bat zeroing in for the kill, it produces its own ultrasonic clicks that interfere with the bat's sonar, leaving it confused. There are also moths that, on hearing an approaching bat, take a serious evasive maneuver and tumble out of the sky.

< Fooled by the echoes coming from the long tails of a moon moth, a bat will end up with a mouthful of these rather than a plump abdomen.

> The cricket fly has perhaps the most sensitive ears of any animal, which enable it to pinpoint its host with unerring accuracy.

The Sound of Silence

In the world of sound, silence is invisibility and the male wax moth (*Aphomia sociella*) puts this to good effect. To attract a mate, he belts out an ultrasonic courtship song, but bats are eavesdropping and will make short work of him if they pinpoint the source. To hide, he simply goes deathly quiet when he hears the hunting clicks of an approaching bat. Crickets and katydids (bush crickets) are also thought to do a similar thing to avoid being nobbled by bats.

The cricket fly (*Ormia ochracea*) has supremely sensitive ears and locates its prey, male crickets, by listening for their song. However, some male crickets (*Teleogryllus oceanicus*) have evolved silence, losing the ability to make their song in the presence of the parasitoid fly. This is evolution before our very eyes—the change from singing to silent males happened in a mere twenty cricket generations.

∨ Bush crickets are also on the menu for bats. The cricket stops singing as soon as it hears the hunting clicks of an approaching bat.

Masters of Camouflage

The next level up from simply hiding to avoid detection is to pretend to be something else and insects are experts in camouflage. Just consider the stone grasshoppers (*Trachypetrella* spp.) that live in arid places strewn with quartzite pebbles in southern Africa. They're so pebble-like that you'd be hard-pressed to see one even if someone pointed it out. Similarly, the moss-mimic stick insect (*Trychopeplus laciniatus*) and the lichen katydid (*Markia hystrix*) can more or less disappear against the right background. The stick insect even has fake tufts of moss adorning its body. Even stick insects with a standard issue, twig-like appearance can be hard to spot.

Other katydids, grasshoppers, moths, butterflies, and treehoppers are also experts at pretending to be plants. Katydids mimic leaves and bark with aplomb. Leaf-mimicking katydids embellish their disguises with pretend fungal spots and fake chewed areas. Katydids, as well as many moths and butterflies, also pretend to be dead leaves to blend in amongst the leaf litter on the forest floor or among dead, withered leaves still attached to plants. The use of browns and contrasting deep, matt blacks in some of the species provides a brain-twisting illusion of rolled dead leaves.

Adult moths are rarely seen without using some manner of light to attract them, in which case we only see them against an unnatural background. Exactly where these animals spend their time when hiding is unknown in many cases. A species of moth (*Calleremites subornata*) I found in northern Myanmar has only been seen a couple of times since it was described in 1894. Against the white sheet of an insect trap, its glorious greens are a real eye magnet, but in its normal daytime resting place it probably disappears against the background—whatever that is—perhaps the underside of a forest leaf. This moth is pretty special, but one of my favorite plant disguises among the insects is that of a Madagascan weevil (*Lithinus rufopenicillatus*), which cultivates a moss garden on its back allowing it to blend in almost seamlessly on moss-covered trees.

< Stone grasshoppers live in arid, rocky areas and they blend right in to the background (*Trachypetrella anderssonii*).

> Top: The weevil *Lithinus rufopenicillatus*, cultivates a moss garden on its body for the purposes of camouflage.

Bottom: Lichen caterpillars blend in almost seamlessly on lichen covered tree trunks (*Enispa* sp.).

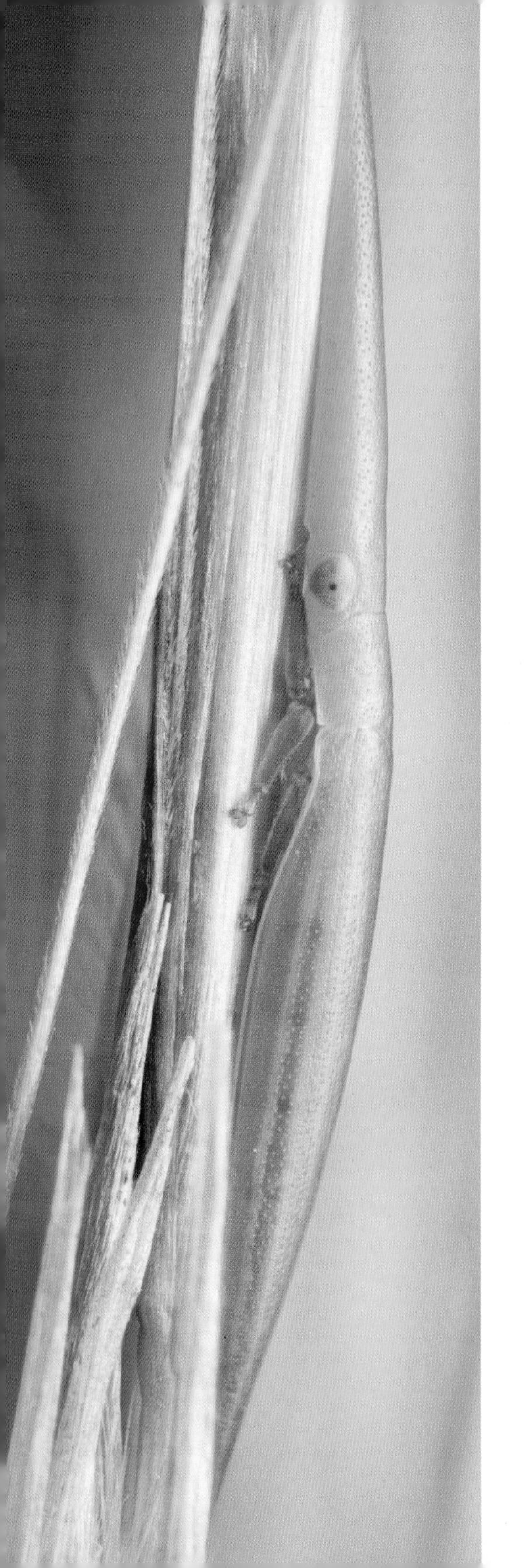

Colorful Disguises

To our eyes, the metallic colors of insects can stick out like a beacon, but it's not us they're trying to avoid. These reflective, often bright surfaces may be difficult for predators such as birds to spot, helping them to blend in on plants, but the exact function of these colors and stylish finishes, if they have any at all, are unknown.

Among the beetles and flies, metallic greens are common and perhaps they give the illusion of a drop of water when the insect is sitting on vegetation. Perhaps most difficult to explain of all are the rainbow hues of jewel beetles. In some cases, these broadcast noxious chemicals that predators would be wise to avoid, but in other cases the glorious colors and patterns of these animals may break up the outline of the body and give it some degree of protection when sitting on sun-dappled tree trunks and branches.

Chemical Mimicry

Disguises, like hiding, are not just about vision. We tend to forget this, as vision is our primary sense, but in the insect world other channels are equally as important, if not more so. This is an area of research where scientists have only really scratched the surface, and most of what happens in the insect world of chemical mimicry is still to be understood. A tantalizing glimpse of what happens there was gleaned from looking at the life and times of the beewolf. The well-stocked nest of this insect (see Beewolf, page 55–57) is a target for other insects that can't be bothered to make and provision a nest of their own, much like cuckoos in the bird world. This is a risky strategy as the beewolf is not to be messed with and any plunderers found in its nest may come to a sticky end. The cuckoo wasp (*Hedychrum rutilans*) has a cunning ploy. It waits until the

^ Bright metallic colors are very common among the insects. Often conspicuous to us, they might be camouflaged to birds (*Eurhinus* sp.).

^ Celyphid flies do a convincing impression of chemically defended leaf beetles, complete with fake elytra.

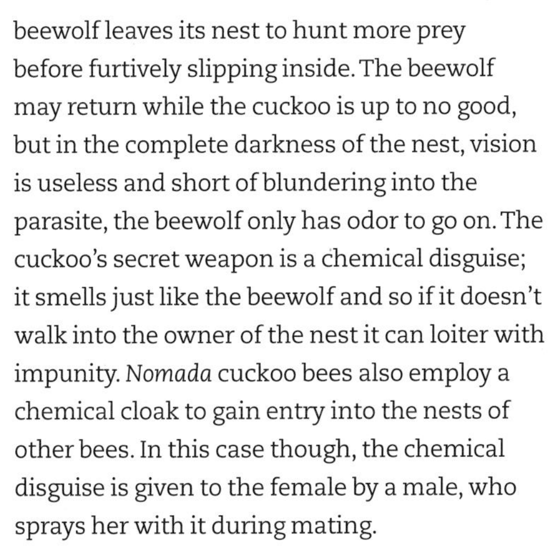

beewolf leaves its nest to hunt more prey before furtively slipping inside. The beewolf may return while the cuckoo is up to no good, but in the complete darkness of the nest, vision is useless and short of blundering into the parasite, the beewolf only has odor to go on. The cuckoo's secret weapon is a chemical disguise; it smells just like the beewolf and so if it doesn't walk into the owner of the nest it can loiter with impunity. *Nomada* cuckoo bees also employ a chemical cloak to gain entry into the nests of other bees. In this case though, the chemical disguise is given to the female by a male, who sprays her with it during mating.

Acoustic Mimicry

As well as odors, sounds can also be copied for the purposes of protection. As well as bright and bold colors, moths can advertise their toxicity to hungry predators with sound, typically ultrasonic clicks. These warning clicks are meant for bats and there are some completely palatable moths that make all the right noises, pretending to be the moths that are loaded with toxins. Acoustic mimicry is probably quite common in insects, especially in those places and times where sound is especially important, such as in dense forest environments and among nocturnal insects.

< The superb camouflage of the leafhopper *Paradorydium menalus*, which lives among reeds.

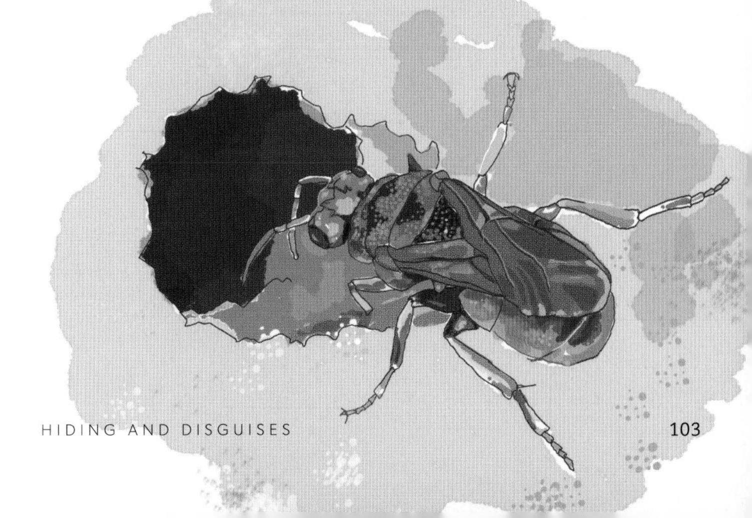

> The cuckoo wasp *Hedychrum rutilans* mimics the smell of its host—the European beewolf (*Philanthus triangulum*).

Fecal Mimicry

Insects have also turned to turds as a neat way of dodging predators. Some smear themselves with their own excrement, some use the excrement of others, others more simply look like feces. In any case, looking like poo or smearing yourself with the real thing is an excellent defense as it repulses predators and can also mask smells that might give you away. Lots of caterpillars have embraced this form of mimicry, specifically pretending to be a small, unappetizing lump of bird poo. It's no coincidence that the main predators of these caterpillars are birds and one thing a bird will turn up its beak at is bird poo. Plenty of adult moths also go for the old bird poo ruse, too, and at least one of them has added something of a flourish—a pair of pretend flies feeding on the pretend poo—or at least this is how it looks to our eyes.

In turn, caterpillar turds, albeit big ones, are the inspiration for the disguise of some tiny leaf beetles and treehoppers. Their surface texturing and ability to fold their legs and antennae out of sight make these insects dropping-doppelgangers, and you'd be hard-pressed to separate them from the real thing, for as long as they keep still.

The subtle camouflage of these turd mimics is very impressive, but why look like excrement when you can just smear yourself in the stuff? This is the strategy of many insects. Take the tortoise beetles, the larvae of which brandish a shield of feces from their rear end, which they use to parry and block the advances of parasitoid wasps and other enemies. This is the height of sophistication compared with the diverse insect larvae that simply coat themselves in fecal matter, such as lily beetle larvae and the weevil *Eucoeliodes mirabilis*. In the latter, Velcro-like hairs

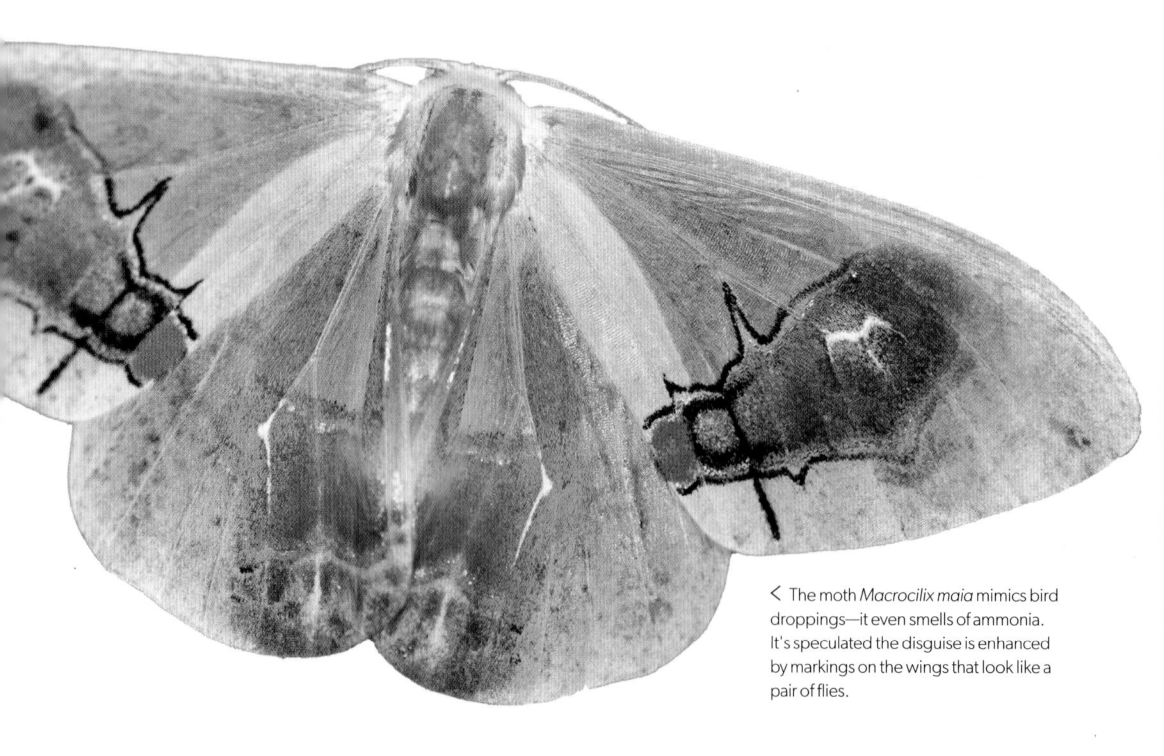

< The moth *Macrocilix maia* mimics bird droppings—it even smells of ammonia. It's speculated the disguise is enhanced by markings on the wings that look like a pair of flies.

keep the feces in place and there are even special adaptations to stop the larva's breathing holes getting clogged with muck.

Waste in all its forms has been embraced by insects to ward off enemies. Lacewing larvae and assassin bug nymphs fashion a macabre suit from the remains of their enemies, which are often trimmed with bits of detritus. The remarkable bone-house wasp (*Deuteragenia ossarium*), a species only described in 2014 from Zhejiang Province in China, makes its nest in hollow stems and similar cavities, which it stocks with spiders as food for its offspring. These provisions and the developing wasp grubs attract predators, so to give her young a fighting chance she tops off the nest with a helping of dead, noxious ants. The residue of noxious chemicals on the ants in this tomb is enough to repel some predators.

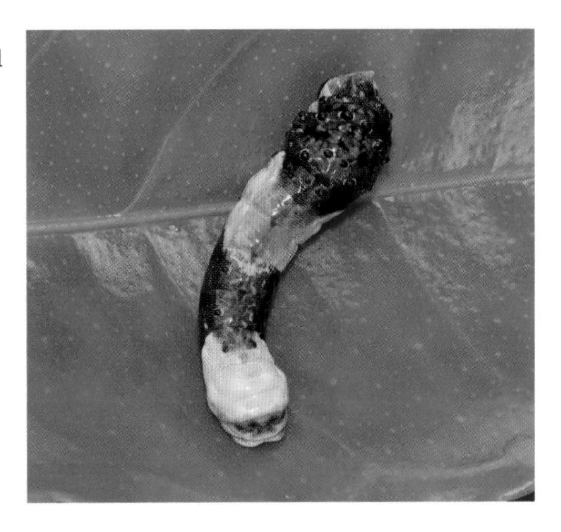

^ Many insects, especially moth and butterfly caterpillars are superb mimics of bird droppings.

v A tortoise beetle larva (R) defends itself with a shield composed of feces and shed skins. The adults (L) are well camouflaged on leaves and can also grip on tenaciously, withdrawing their legs and antennae underneath their flanged pronotum and elytra.

∧ Eyespots are common in the insects, especially among the butterflies and moths. It is thought these deter predators by giving the impression of a much larger animal.

∨ The markings on the folded wings of the lacewing *Psychopsis mimica* appear to resemble a spider.

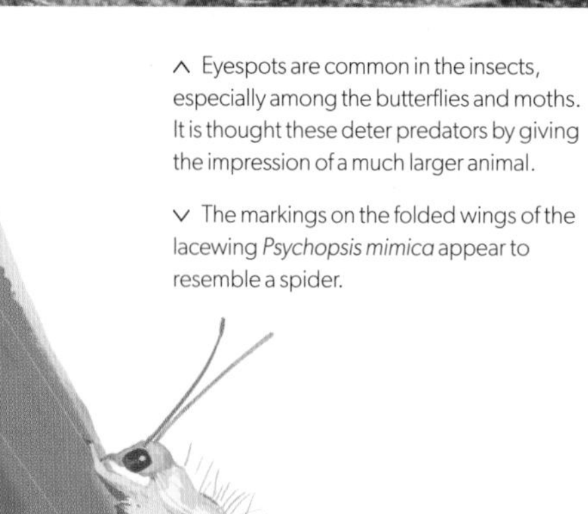

Mimicking Other Animals

As well as using waste and pretending to be inanimate objects, plenty of insects also appear to mimic other animals for the purposes of protection. Some large caterpillars can do a convincing impression of a snake, lots pretend to be ants, and others try and channel their inner spider. There's an unusual lacewing *Psychopsis mimica* that has the spectral image of a spider on its wings, complete with eyes, well that's how it looks to us humans. Young true bugs and crickets ape ants because ants are well protected and there's often lots of them; therefore, many predators steer well clear of them.

There are even flies that mimic leaf beetles, planthoppers that mimic weevils, weevils that mimic flies, flies that appear to mimic spiders, cockroaches that mimic ladybirds, and an abundance of insects from various orders that mimic wasps. The reason for mimicking a wasp is clear-cut, wasps have a painful sting, so lots of predators keep away from them. Another well used ruse is that of the false head. Pretending your back end is your head will hopefully direct the attentions of a predator to a part of your body that can be sacrificed.

Batesian and Müllerian Mimicry

When an animal that tastes good or is undefended pretends to be an unpalatable or otherwise well-protected species, this is known as Batesian mimicry. Familiar examples of this are hoverflies pretending to be wasps.

Müllerian mimicry is where two or more species that taste disgusting or are defended in another way and which share the same predators evolve to mimic each other. Over time, other species may join in, too, until there's what's known as a mimicry ring. Mimicry is rarely clear-cut though, as cheats can evolve and what our eyes see may not be what is seen by the eyes of other animals.

> Wasps can sting, so lots of other insects mimic them. This fly (*Monoceromyia* sp.) mimics a potter wasp.

∨ This clearwing moth is a convincing mimic of a wasp, even down to the way it flies (*Similipepsis* sp.).

The Eye of the Beholder

What we take to be mimicry is sometimes not the case. Just because we see something one way doesn't mean that the eyes of other animals see it the same way. For example, the fuzzy little velvet ants (*Dasymutilla gloriosa*) (not ants and definitely not velvet) of the arid, southwestern United States, were long assumed to be mimicking the fuzzy seeds of the creosote bush. But, it turns out their appearance is more about staying cool than fooling predators. Their long, white hairs reflect heat allowing them to scuttle around on the scorching sand looking for the burrows of sand wasps—their hosts. Their similarity to the seeds of the creosote bush may be convergent evolution—the fuzzy covering of the seeds may also be about staying cool.

∧ The fuzzy velvet ant *Dasymutilla gloriosa* was once thought to mimic the seeds of the creosote bush, but their appearance is more about keeping cool.

∨ The young of many bush cricket species are thought to mimic ants.

TAKING FLIGHT

When or even before an insect's hiding place gets discovered another line of defense is simply to make a sharp exit. Flying is the obvious way in which insects can avoid danger, but many insects also have prodigious jumping abilities.

Try to grab a flea, a flea beetle, or a planthopper and they take off like a rocket, propelled by enormous leg muscles and a special type of protein called resilin, which is extremely elastic. These prodigious leaps can sometimes seem a little uncontrolled, so some leafhoppers have evolved gears. These keep their legs completely together allowing them to make more measured jumps.

My personal favorite is the escape tactic of the *Stenus* rove beetles. These beetles are truly blessed in bizarre adaptations. If telescopic mouthparts weren't enough (see page 84), they also have one of the most impressive defenses of any animal. Often, these beetles are found in waterside habitats, where they wander about on the ground and in the vegetation shooting their mouthparts out at mites and springtails. Their small size and long legs means they can also take to the water and scull along as if they were walking—their weight supported by the surface tension.

If a predator threatens while they're on the water, these beetles emit a chemical (stenusin) from the glands at the tip of their abdomen. This substance is so hydrophobic that as it hits the water it breaks the surface tension, propelling the beetle forward at terrific speed, well, relatively. The velocity the beetle reaches is equivalent to about 370–560 miles per hour (600–900 km/h) in human terms and it does this in a fraction of a second; this acceleration would turn us humans inside out. Needless to say, the sudden disappearance of its dinner will completely bamboozle the predator.

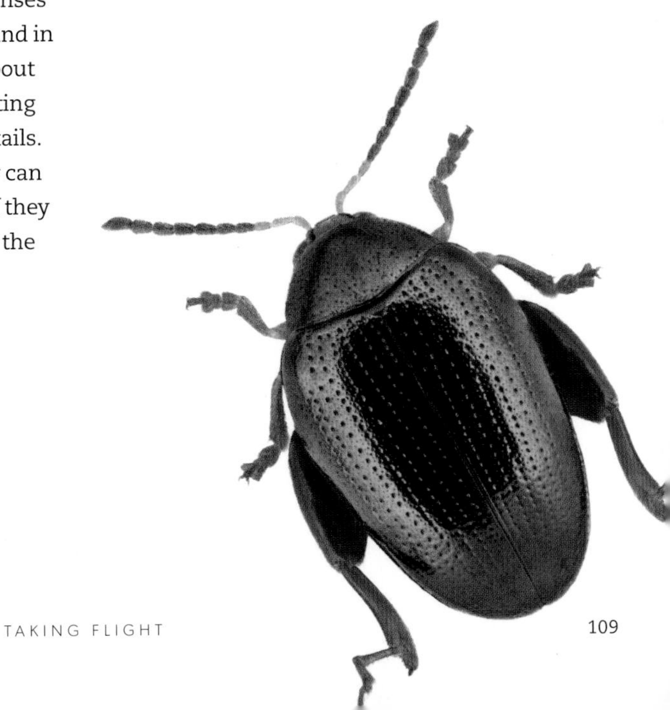

> Using their relatively enormous, muscular hind legs, flea beetles are prodigious jumpers (*Psylliodes chalcomera*).

∨ *Stenus* rove beetles can skim across the surface of water at great speed by releasing a surfactant from their anal glands. This chemical disrupts the meniscus, propelling them forward.

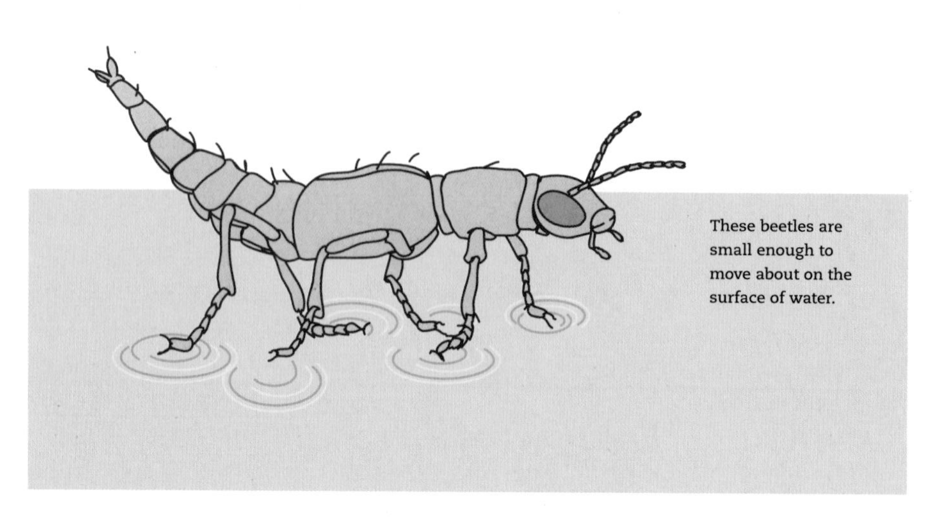

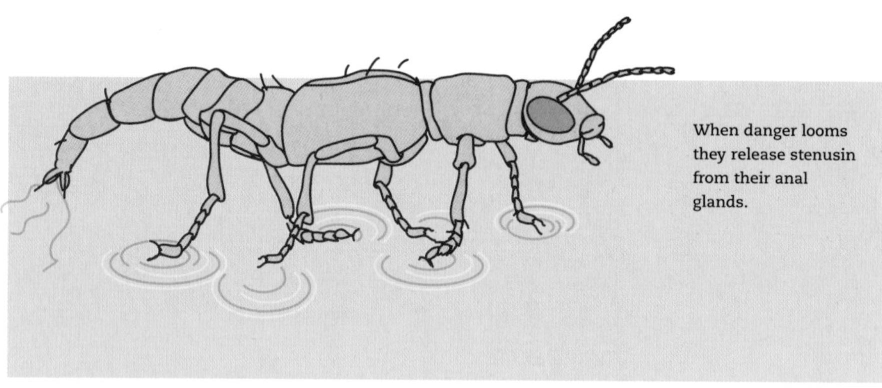

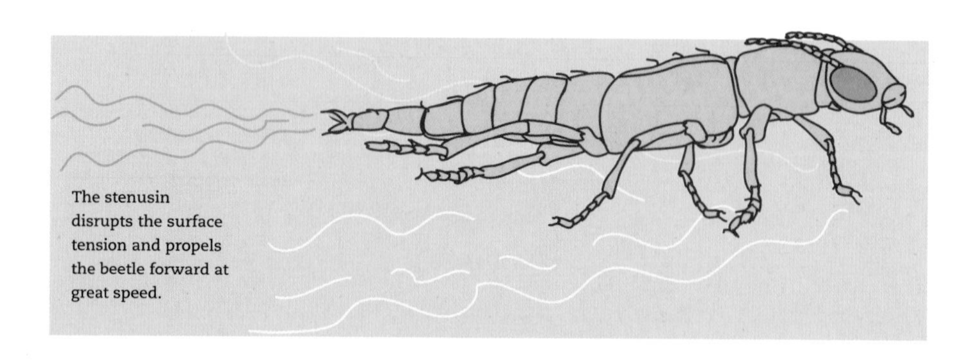

INSECT ARMOR

The exoskeleton is a real marvel of evolution. It gives an insect mechanical protection and prevents water loss. With a few tweaks to its chemical composition, it can be soft and pliable or enormously tough.

The exoskeleton of the aptly named ironclad beetles (*Zopherus nodulosus*) is so tough that it can bear the weight of an adult person and an entomologist's pin bends against it. Weevils are similarly furnished with a very tough exoskeleton, which is probably one of the reasons for the success of these animals, as there are lots and lots of weevils. Beyond simple toughness of the exoskeleton, time and danger has molded into it all manner of spines and spikes that are primarily a defense against vertebrate predators. Some beetles have gone particularly medieval as they bristle with cruel-looking spikes that are just begging to get snagged in the craw of a hungry bird or

∨ Some weevils are among the most heavily armored insects. They can also play dead, folding their snout, antennae, and limbs into shallow recesses (*Cratosomus* sp.).

^ Some midge larvae can produce waxy balloons. These are waggled as the larva moves and they fall off easily, which suggests they are a distraction for predators, but no one knows for sure (*Forcipomyia* spp.).

mammal. Lots of caterpillars also defend themselves with spines and many have the added advantage of being loaded with venom, so when they break off in the skin of a predator they can cause anything from discomfort all the way up to life-threatening allergic reactions.

In addition to all these spikes, there are mandibles and claws. In some cases, these are impregnated with metal ions to allow them to hold a keen edge for longer. Some of the beetles, crickets and their relatives, wasps, ants, and termites have particularly powerful mandibles that are more than capable of delivering a nasty nip to potential predators. Some of the longhorn beetles have mandibles akin to bolt cutters, which I've been on the receiving end of several times. They can slice straight through human skin. The insects endowed with these serious mandibles often hold them agape as part of their warning display if they feel threatened—almost as an invitation to a predator that is desperate enough to try its luck.

Plump and juicy insect larvae are particularly at risk from predators, so this is where we find the ability to make all sorts of neat little cases, from the extraordinary creations of caddisfly larvae to the casks of excrement fashioned by various beetle larvae. Caddisfly cases are something to behold and depending on the species all sorts of building materials are used, from pieces of plant matter to the empty shells of tiny snails and grains of sand. In all cases, the building blocks are bound with silk and assembled with incredible precision.

Silk has also been harnessed by lots of other insects for the purposes of protection. Moths wraps themselves in it for the delicate process of pupation, sometimes reinforcing it with wood and plant fibers to make an extremely

∧ The fluffy, probably defensive ornaments of another tiny *Forcipomyia* midge larva.

∨ This bag worm moth has just emerged from the protective case it carried around as a caterpillar.

tough cocoon. The web spinners, slinky, easily overlooked insects can spin enormous amounts of silk from glands on their forelegs to make the labyrinthine silken tunnels in which they live. The silk is their main defense against predators and it also maintains the perfect microclimate for these enigmatic insects.

Moths and butterflies are rather flimsy animals, but the scales on their bodies and wings are an ingenious defense against all sorts of enemies. If one of these animals flutters into a spider's web, the scales get stuck rather than the owner, and it can simply peel itself off and continue on its way none the worse for wear apart from missing a few scales. Even more elaborate is a tiger moth (*Homoeocera albizonata*) from South America that churns out loads of fluff when threatened, probably enough to deter or even confuse bats and other predators besides.

What with their tough exoskeleton, all is not lost, even when an insect is swallowed by a predator. After being gulped down by a hungry frog, one tough species of water beetle (*Regimbartia attenuata*) scuttles through the predator's gut straight for the back door and gets out with the frog's waste. It might even tickle the amphibian's sphincter to hasten its escape.

∨ Many true bugs secrete waxy filaments/coverings. These are thought to be a distraction/defense against predators and parasitoids (*Nogodina* sp.).

CHEMICAL WEAPONS

Insects are particularly successful in the use of chemicals to defend themselves. The substances produced by these animals in the name of protection are incredibly complex. In some cases, they're produced by the insect, others are pinched from plants, and yet more are the work of symbiotic microorganisms that live inside the insect.

The chemicals produced by insects might ooze from the limbs or mouth, be squirted or bubble from special nozzles, or be wafted under the quivering snout of an attacker. They vary from offensive odors all the way up to extremely potent toxins.

Plant and fungal defensive chemicals are even more diverse than those of insects, and this has been exploited to the full by these six-legged marvels. The bulk of these defensive chemicals are produced to deter plant and fungal feeding insects, but in an evolutionary arms race that has been raging for tens of millions of years, the insects have evolved to not only eat these toxin-laden tissues with no ill effects, but also harness the chemicals for their own defenses. Sometimes, the toxins are tweaked by the insect to make them even more unpleasant.

∨ Insects that are chemically defended broadcast this to potential predators with bright colors and bold patterns. This is known as aposematism. This is also the basis of Batesian mimicry, where harmless species evolve to resemble harmful ones (*Harroweria* sp. Bush cricket nymph).

The list of insects that do this would fill a book and familiar examples of this chemical theft are all around us. Take the various insects that feed on the foliage of milkweed. Milkweed is laden with some potent toxins, which do nothing to deter these specialist herbivores, who munch the foliage with impunity and tank themselves up with the toxins. Only a naive, deranged, or ravenously hungry predator would attempt to eat these toxin-pumped herbivores.

These well-protected insects need to broadcast their defenses, otherwise they run the risk of being caught and then promptly spat out by a hasty predator. Bold bright colors, sounds, and odors are all used to advertise these toxins. Just think of the red and black of a ladybird, or the yellow and black of a social wasp. This aposematic coloration is a deliberate warning to repel enemies. Bright colors are often paired with dramatic eyespots, which are used as part of a startle display to give the predator a fright and hopefully distract it. For full effect, otherwise camouflaged insects flash concealed bright colors in a dramatic startle display if they've been discovered.

Repellent Smells

The chemical defenses of some insects are all about repulsion, which include rank odors that make your eyes water. Stink bugs have a peculiar, very strong aroma, which sticks to your skin for a long time after you've handled them. The secretions of giant stink bugs are potent enough to cause painful burns, which I've experienced when holding one too close to my nose. Have a casual sniff of the back end of a carrion beetle or the front end of a shore earwig and you'll quickly regret it, as your nose will be assaulted by a cloying, fetid stench that takes some shifting. Smelling like death and decay is a surefire way of putting some predators right off their dinner.

Defensive Chemicals from Symbionts

Rather than producing chemical defenses themselves, lots of insects have an arrangement with microbes that pump out the toxins in exchange for food and shelter. *Paederus* are conspicuous rove beetles, often found near water. Their bold colors broadcast that they are loaded with the toxin pederin. Get some of this on your skin and you'll develop painful sores and welts. This toxin is produced by a *Pseudomonas* bacteria that lives inside the beetle. Symbiotic microbes may be behind the chemical defenses of many insects, especially those that don't feed on plants or fungi.

Venomous Defense

Venoms are also well used for defensive purposes. Assassin bugs, even the small ones, have a very painful, venomous bite that is enough to make you think twice before handling them. The stings of wasps and bees pack a real punch with a venomous sting that is all about leaving a lasting impression by causing pain and triggering immune responses in potential enemies. In the social wasps, bees, and ants the threat of hundreds or thousands of workers each with a painful sting is enough to deter even large animals. Even a longhorn beetle has jumped on the venom bandwagon with the tip of each antenna modified into a venom-loaded stinger that can be jabbed into a predator.

< The bold, bright warning colors of an Amazonian wasp moth, broadcast its toxicity to potential predators. *Isanthrene* sp. wasp moth.

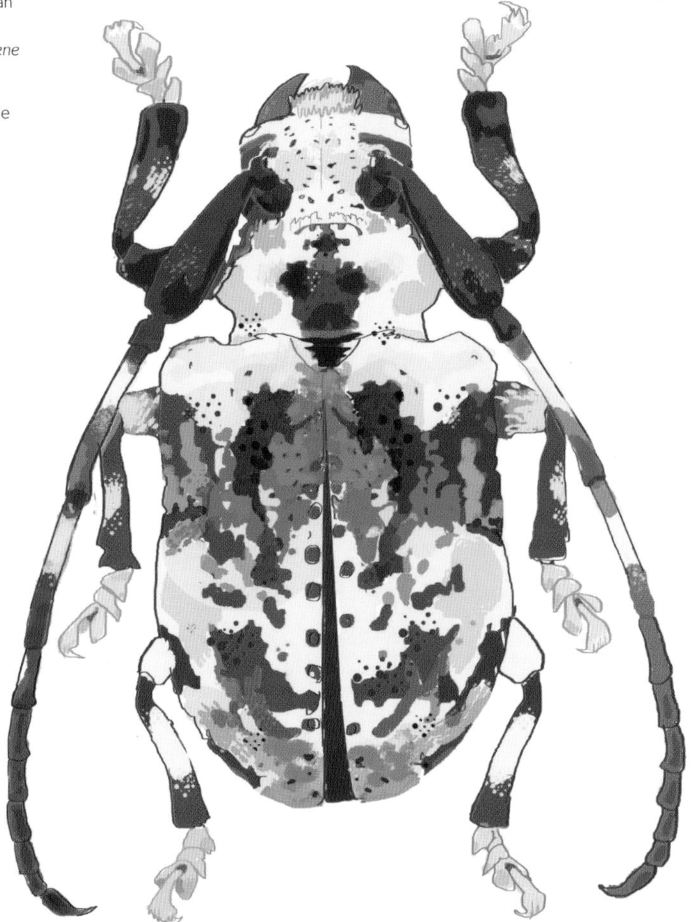

> To defend itself, this longhorn beetle can sting using the tips of its antennae (*Onychocerus albitarsis*).

BOMBARDIER BEETLES

There are about 500 species of bombardier beetle and they have arguably the most impressive chemical defenses of any insect. Glands in the abdomen of these insects have evolved to produce compounds that are stable when in isolation, but react with alarming ferocity when mixed together. A reaction chamber at the back end of the beetle allows for the mixing of these chemicals and the subsequent pulsed release of boiling hot vapor.

I've handled some of the larger tropical bombardier beetles and the hot, caustic vapors make you drop them like a hot potato and leave your fingers stained a purplish brown. They're also very good at aiming their explosive weapon. Imagine when this boiling, noxious mixture explodes into the face of a small mammal or bird. It must be excruciating. It's no surprise that these beetles have few predators. Nature always finds a way, though. Toads appear to be stupid enough to wolf down bombardier beetles, but in some cases they vomit them up later after the beetle has let off some explosions inside their belly. Some spiders can subdue these explosive characters by quickly wrapping them in silk to absorb the hot, mephitic cloud.

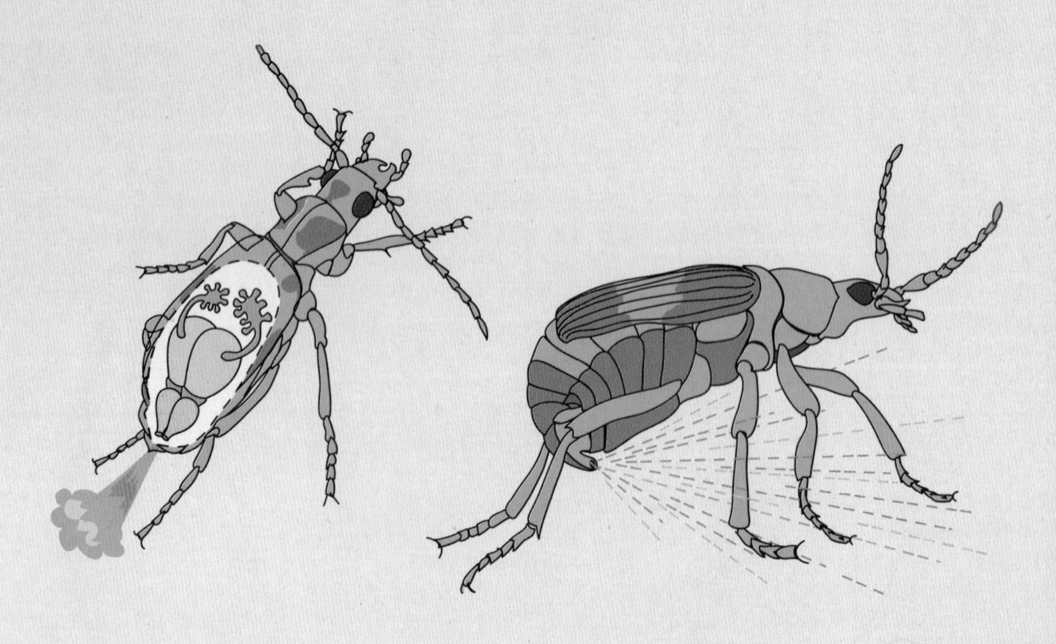

∨ Bombardier beetles have perhaps the most elaborate and effective defense of any insect. Chemicals from separate glands are brought together in a reaction chamber with explosive results.

Counteradaptations

The dizzying array of defenses displayed by insects are perfect examples of the non-stop struggles that exist in the natural world. Any adaptation that evolves to counter a predator is quickly met by counteradaptations that give the predators the upper hand for a while, until the downtrodden evolve something better. A nighttime encounter in the rain forest of Peru that reminds me of this was finding the disembodied, but still living head and thorax of a big longhorn beetle crawling hopelessly on the trail. Some unknown predator in the trees above had somehow yanked the heavily armed front from the juicy, relatively soft abdomen and cast it aside, completely sidestepping this unfortunate insect's defenses. To some people, the ceaseless struggle between predator and prey can appear brutal, but they should remind us that nature is vibrant and ever changing.

∧ When threatened this large, conspicuous fly ejects a bright yellow, noxious fluid from its mouth (*Bromophila caffra*).

∨ Any predator that tries to eat an oil beetle will get a mouthful of toxin laden hemolymph that oozes out from between the segments of its limbs. This is known as reflex bleeding and the toxin in question is cantharidin.

4

SOCIALITY

Lots of animals live in groups, such as nesting aggregations of birds, shoals of fish, and dense clusters of anemones on the seabed. To find the zenith of group living among animals, we must look at insects because this is where we see eusociality. This is an advanced form of group living where a single female or small numbers of females have offspring while nonreproductive individuals, sometimes huge numbers of them, do everything else for the maintenance of the colony, such as childcare, homemaking, and providing food.

A RARE WAY OF LIFE

Although eusocial insects are everywhere—just think of how many times in your life you've seen ants—this way of living is actually quite rare across the insect tree of life. It is mostly restricted to termites, wasps, ants, and bees, although some beetles, aphids, and thrips have adopted this lifestyle.

Eusocial insects such as ants, honey bees, and termites are conspicuous and familiar animals. Ants have to be among the most easily recognized animals on the planet. Some of these insects, such as honey bees, are also the most studied of all the animals.

Yet, because of this familiarity few of us ponder the strangeness of these eusocial insects, and we've barely scraped the surface of understanding them (see Control of the Colony, page 125). Some species can be studied with relative ease. Just think of a commercial honey bee hive and how easy it is to see what's going on inside the colony. The nests of most other eusocial insects though are like a black box, and understanding what's going on inside is impossible without disturbing them in some way.

< Social wasps at their nest (*Ropalidia* sp.).

∨ The individuals in a eusocial insect nest are in near constant communication with one another.

^ In social insects there is a whole spectrum of organization, from primitive to advanced. Stenogastrine wasps are in the former camp with nests that usually contain less than ten adults.

One of the strangest things about these eusocial animals is that nearly all of the individuals in these colonies have given up on reproduction, instead devoting their efforts to building and maintaining a nest, and supporting a single or small number of queens that can reproduce. In the light of natural selection, the life of a single ant, a social wasp, or a termite, doesn't make a lot of sense, but when we place this one insect in the context of the colony, it does make sense. In many of these eusocial insects, the colony "works" because of the very peculiar way in which the individuals start life.

This fascinating way of life is restricted to a few groups of insects. The termites, effectively wood-munching cockroaches, are all eusocial. In the Hymenoptera (wasps, ants, and bees), this way of life has evolved at least eleven times. Nearly all the ant species are eusocial, except for a few slave-making species that have no workers and rely on labor supplied by workers of their host ants.

We tend to assume that all bees and wasps live in some kind of nest or hive where they clamor around a single queen. The truth is that most bee and wasp species are solitary. It's important to remember that both the ants and the bees evolved from solitary, predatory wasps. Beyond the termites and the eusocial Hymenoptera, which are far and away the most significant eusocial animals, this way of life also crops up in the aphids, thrips (see Social Aphids and Thrips, pages 126–127), and beetles. As well as the animals above, there are other insects that appear to be on the verge of eusociality.

Control of the Colony

How does a colony of social insects, thousands or millions strong, actually work? This has to be one of the most fascinating questions in all of nature and like so much else, we don't really know. The old view was that top-down instructions from the queen in the form of chemicals controlled every aspect of the colony. This was based on how human societies work—a hierarchy with the elite controlling the underlings. We now know this is not the case and that these colonies are probably stranger than we can imagine. Much about the lives of these eusocial colonies only makes sense if we view the entire colony as the "individual"—a kind of superorganism. In the superorganism, the queen is the colony's ovary. The nurse workers are reproductive accessories, tending to the brood and the queen. The foragers are the "muscle," responsible for all the other things that keep the colony working. In the superorganism, no one individual is in control and no one individual can ever know the colony's needs. For example, a single leafcutter ant carrying a piece of leaf back to the nest, or a termite bursting itself in defense of the nest, have no idea what's going on in the colony at large.

What we do know is that the individuals in a colony are always exchanging information. Watch some ants for a while and you will see them seemingly "greet" one another by touching antennae. This is where information via chemical messages is being shared. For example, they might be letting each other know about threats and food sources. Crucially, each of the workers is only following a simple set of rules and responding to what's happening in its local environment. When there are lots of workers all doing this incessantly something remarkable happens—we see the emergence of complex behaviors.

> Colony members often exchange regurgitated liquids. This is known as trophallaxis and ensures the equal sharing of food through the colony.

∨ Only the "queens" in a eusocial insect nest produce offspring.

SOCIAL APHIDS AND THRIPS

Aphids often live in dense aggregations on plants and lots of species also induce their host plant to produce a gall in which they live. There are plenty of species where these aggregations appear to be complex. However, whether or not these groups can be considered to be eusocial is a bone of contention as they are sexless beasts and all the individuals in a given group are clones of the female that started the colony. In any case, there is a division of labor as some individuals are soldiers, who defend the colony from enemies and take care of other tasks, such as removing waste from the colony. These soldiers can be equipped with enlarged legs, horned heads, thickened exoskeletons, and venom-loaded mouthparts. With these weapons they can grab, crush, pierce, and sting their enemies. Some gall-dwelling soldier aphids even sacrifice themselves to protect the colony. When their gall is ruptured by a hungry caterpillar, they first try and deter it by stinging it. They then try and repair the damage to the gall by using their own bodily fluids, which rapidly coagulate to fill the hole. Some soldiers get trapped in the patch as it's hardening, while others get trapped outside. In either case, it means a slow and grisly demise for some of the defenders.

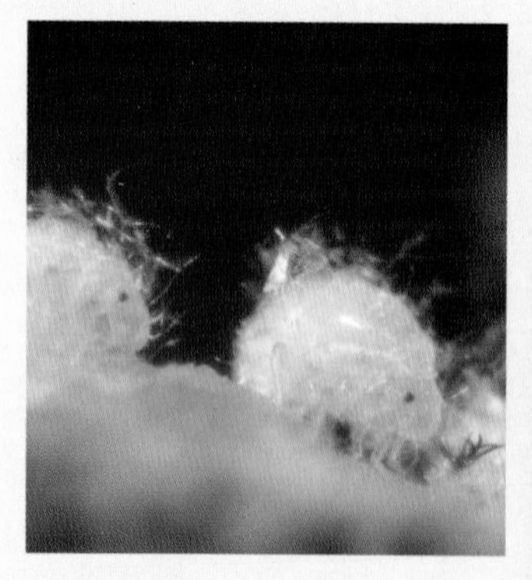

∧ The nymphs of this social aphid species rupture their bodies to seal breaches in their gall (*Nipponaphis monzeni*).

> Social aphids inside their gall (*Nipponaphis monzeni*).

Complex colonies of some Australian thrips also inhabit galls. Some of the offspring of the female who created the gall become soldiers with large forelegs for defending the colony, especially from other thrips that want to take over the gall. The rest of her offspring will become dispersers, which leave the gall to try and found their own colony.

The waters of harmonious, eusocial living are a bit muddied here, as it appears that some of these soldiers may be having offspring of their own under the nose of the queen. It also turns out that the soldiers of at least one species are medics, dispensing anti-fungal compounds to keep pathogenic fungi at bay. All of this by insects barely 0.04 inches (1 mm) long living in a tiny gall on an acacia!

GALL DEFENDERS

- 60 species of soldier-producing aphid are known
- The toxins injected by soldier aphids can paralyze and kill their insect enemies
- A high density of aphids in a gall can be the trigger for the development of soldiers

^ Soldiers of the social aphid *Tuberaphis styraci* attacking a predatory lacewing larva.

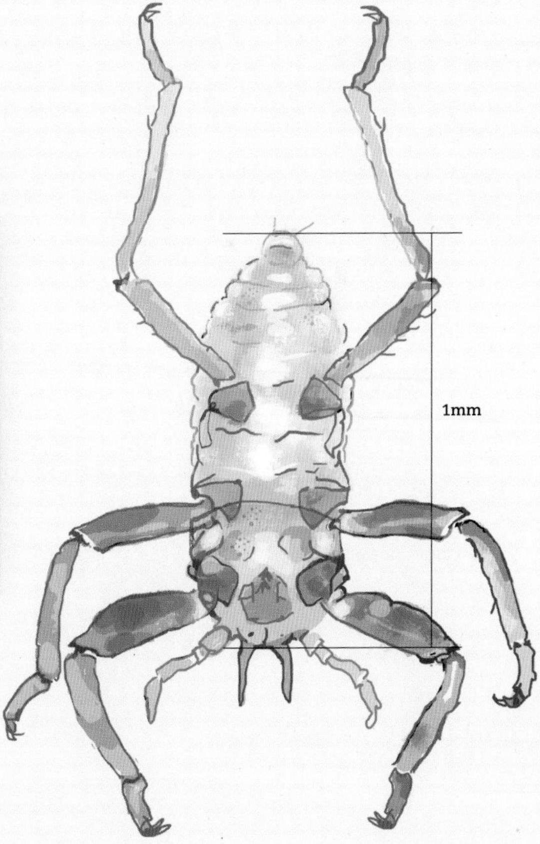

> Some social aphids have aggressive soldiers that jab their frontal spines into predators (*Pseudoregma panicola*).

One Set of Chromosomes or Two

In many of the eusocial insects, establishing a nest depends on the queen being able to produce lots of daughters, who will maintain and expand the nest, tend to the brood, find food, and repel enemies. This ability to produce an army of loyal daughters, hinges on the extraordinary nature of reproduction in these animals. The queen ant can control the sex of her offspring by very sparingly using the sperm she stored from her single mating. If a daughter is needed, the queen releases a tiny amount of sperm to fertilize the egg as it's being laid. If the queen withholds the sperm, the egg is not fertilized and it will develop into a son. This quirk means that a male of one of these insects has a grandfather, but no father and will have grandsons, but no sons. This unusual feature is at the root of the success of wasps, ants, and bees, because it means that a female worker is more closely related to her sisters than her own offspring, so it pays for her to look after her sisters and the nest, rather than strike out on her own and have her own babies.

The ancestors of the living eusocial wasps, ants, and bees were solitary, predatory beasts that stocked their nests with paralyzed prey—as do the living solitary wasps. Often, these "nests" are a simple, linear arrangement of brood cells constructed in the ground, or in a hollow plant stem, or in dead wood. Each brood cell contains an egg, provisions of food (paralyzed prey), and it is separated from the next brood cell by a partition often crafted from mud. The daughter wasps are bigger than the sons, so they need more food and more time to develop. The mother wasp has the ability to lay female eggs in the deeper brood cells of the nest, and to lay male eggs in the brood cells nearer the entrance. Without this quirk, the male offspring would be developing first and causing havoc by trying to escape the nest and breaking through the brood cells of their still-developing sisters.

∨ This type of sex determination is also at play in solitary bees and wasps—the ancestral lifestyle of the social species. This image shows the nest of the Mexican grass-carrying wasp (*Isodontia mexicana*) nest. It shows tree cricket prey, grass blade cell partitions, and a cocoon (center left). Female eggs are laid first in the deeper parts of these linear nests because daughters are bigger than sons and need more time to develop.

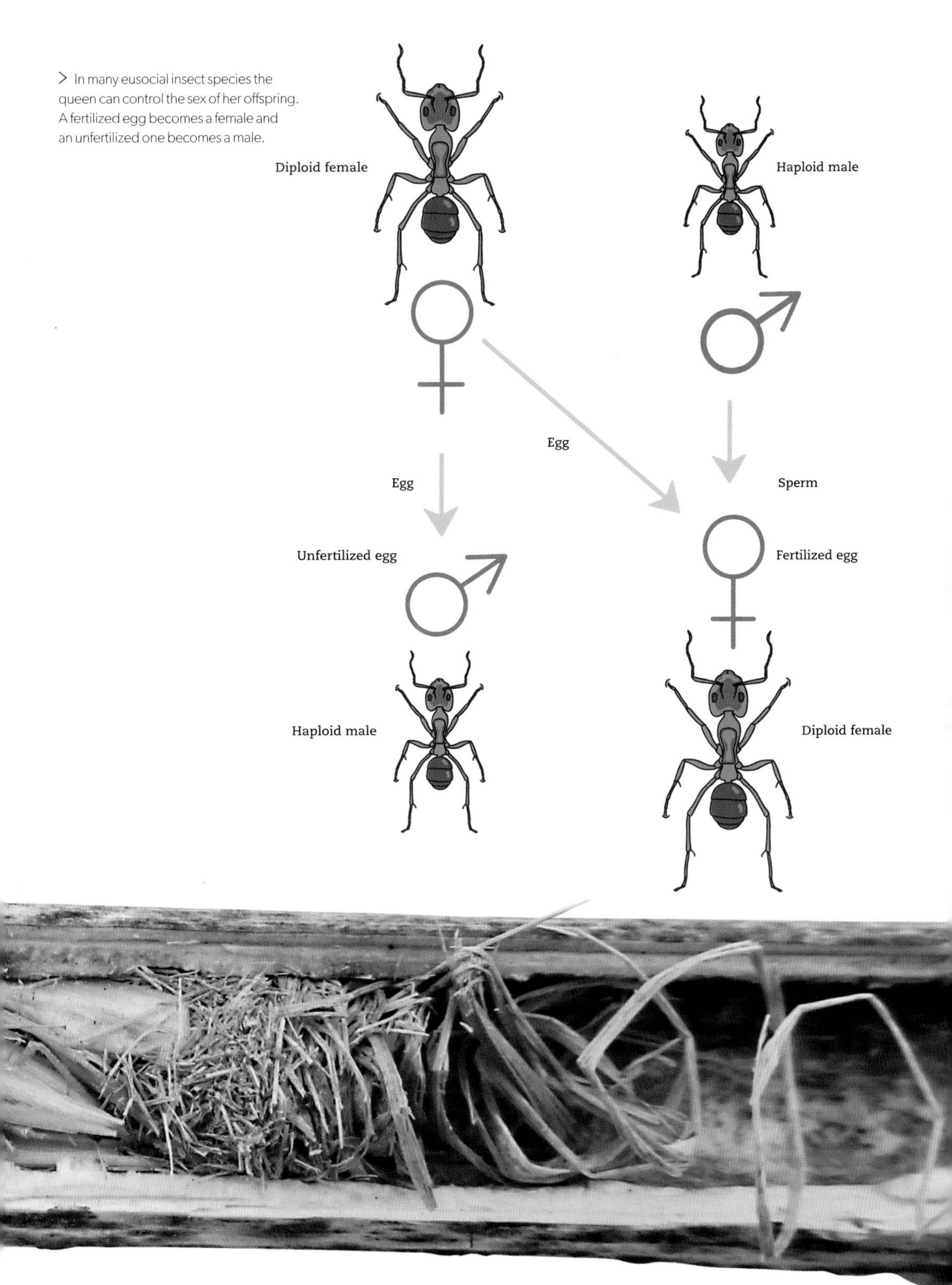

> In many eusocial insect species the queen can control the sex of her offspring. A fertilized egg becomes a female and an unfertilized one becomes a male.

ECOLOGICAL DOMINANCE

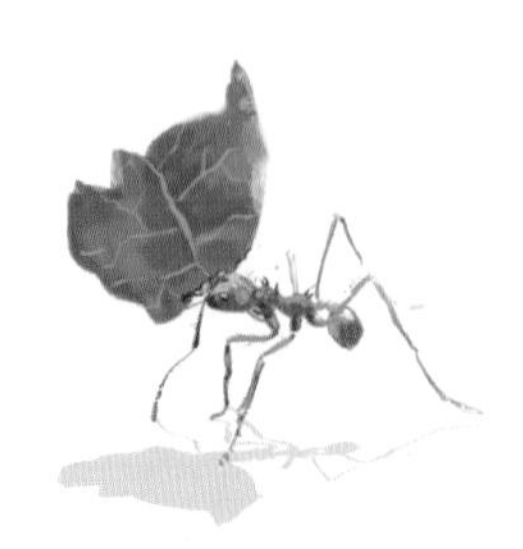

Although not all that many insect species are eusocial the ones that do live this way are dominant players in their ecosystems because of their impressive organization and enormous numbers. In the tropics there are ants and termites everywhere.

In the tropics it's not unusual to find 1,000 ants or termites in any given square yard (meter) of ground. In tropical and subtropical ecosystems, termites are the major eaters of the two most abundant biomolecules on land—cellulose and lignocellulose—essentially the stuff that makes up plants. Collectively, termites eat 50–100 percent of the dead plant biomass in these ecosystems. All of this eating generates a lot of gas and 2–5 percent of all the methane in the planet's atmosphere is thought to be from termite farts and the decay of matter in their nests.

Likewise, leafcutter ants dominant ecosystems in a way that bears no relation to the size of an individual ant. These ants form perhaps the most complex insect societies, building enormous and fantastically complex nests that can extend over 6,458 feet^2 (600 meters^2) and reach down 26 feet (8 meters) or more into the soil. These colossal nests are home to millions of workers.

To get an idea of the size and complexity of these nests, biologists pour cement through the holes on the surface, which then hardens and can be carefully excavated. Filling all the tunnels and chambers can take as much as 11 tons (10 tonnes) of cement, eventually revealing a veritable insect metropolis. The tunnels connecting the chambers are constructed in a way to maximize airflow through the nest and to provide the shortest transport routes.

> In subterranean chambers of leaf-cutter ant nests, fungus gardens are cultured on the plant material. Swellings produced by the fungi are what the ants eat.

∨ In gathering plant matter and constructing huge, complex nests, leafcutter ants have a huge impact on the habitats where they live.

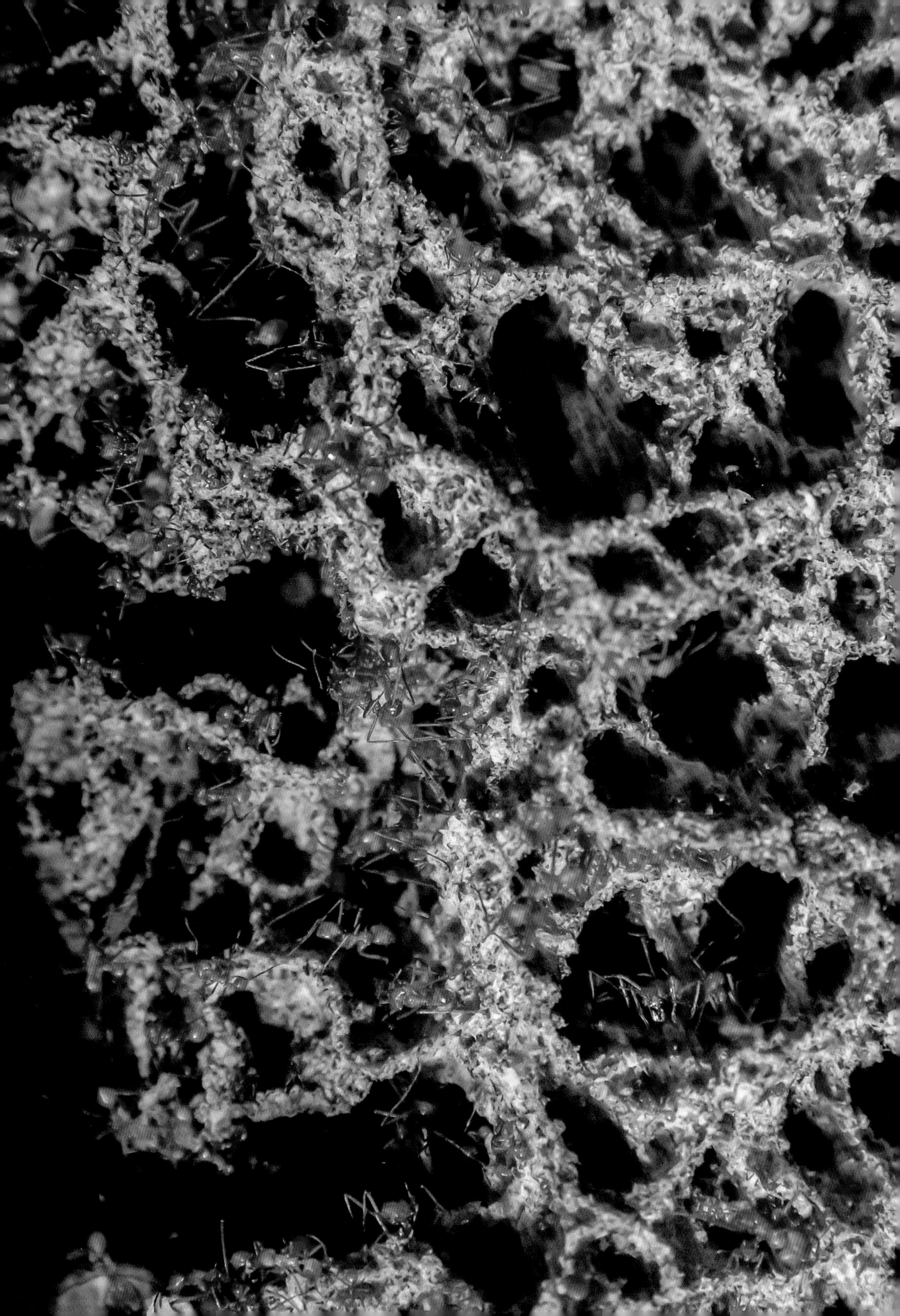

Amazingly all of this work is done by animals with brains about as big as this full stop. In constructing these nests, the ants have to move tons of earth, approximately 44 tons (40 tonnes) for a nest covering only 538 feet² (50 meters²). This Herculean task equates to billions of ant-loads of soil, each load is carried over half a mile (1 km) in human terms, and weighs four times as much as the ant itself. These structures are truly a wonder of the natural world. In stark contrast to those ants that make enormous, rambling nests are those species (e.g. *Temnothorax* sp.) whose entire nest can be contained within a single acorn or similarly tiny space.

The subterranean efforts of the leafcutter ants are matched or even exceeded by what they accomplish above ground. In a strange symbiosis, and one of the few examples of farming among the animals, leafcutter ants grow and eat a type of fungus, specifically little swellings called gongylidia provided by the fungus. The fungus is cultivated in special gardens held within some of the subterranean chambers. The fungus is only found in the nests of leafcutter ants. The fungus in turn consumes plant matter and this is what the ants have to collect, making them among the most important plant-eating animals in the places they're found. Long lines of foraging workers scuttle off into the surrounding landscape, completely defoliating large areas to keep the fungus fed. In some places, leafcutter ants consume more foliage than all the herbivorous mammals combined in that area.

∨ Leafcutter ants' nests are extremely complex and they can be huge.

> To reveal the size and complexity of a leafcutter ant nest, cement can be poured into the entrances and then left to set. The soil is then carefully excavated to reveal the chambers and tunnels.

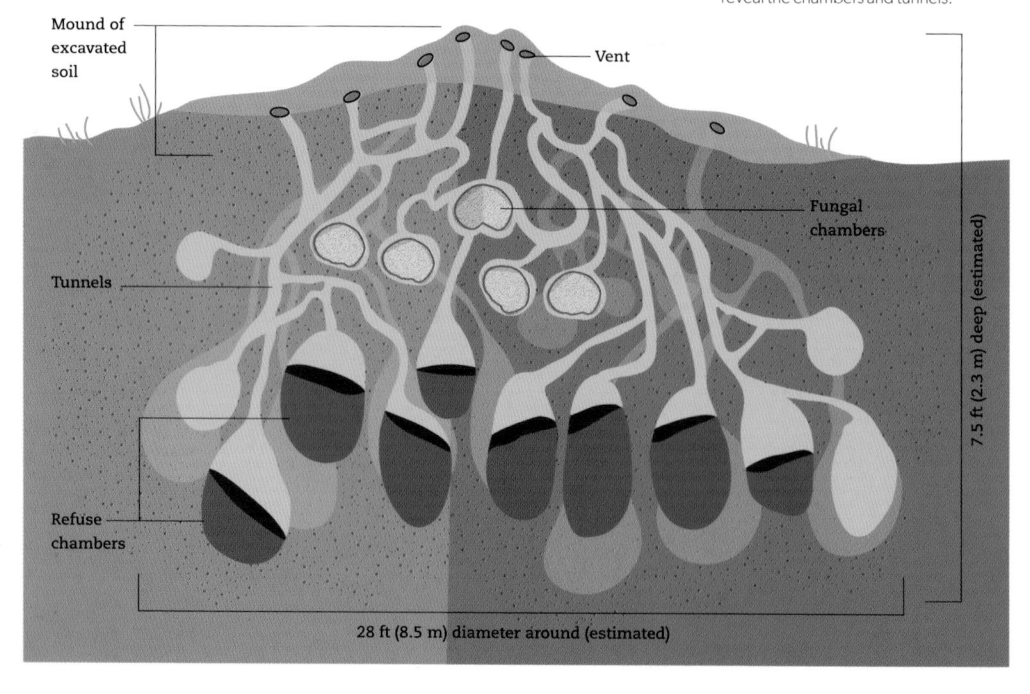

WHAT ANTS EAT

Ants eat all sorts of things. Some are fierce predators that catch and carve up live prey that is carried back to the nest. Wood ants are capable predators and scavengers, but their most important food source, making up 90 percent of their diet, is honeydew—a sweet liquid that oozes from the back-ends of aphids high in the canopy of the trees that surround the nest. Some ants eat little packets of nutrients known as Beltian bodies, which are provided by the leaves of certain trees.

Dracula ants have among the strangest diet of any ant. When food is in short supply the queens of these ants resort to feeding on their own brood—picking up a larva, piercing it gently with her mandibles and drinking the hemolymph. In some species, the workers in the colony will also resort to sucking their siblings, but they do so sneakily by carrying the larvae out of the brood chamber first. This strange behavior doesn't kill the larvae, but it may slow or stunt their growth. The larvae of some species (for example, *Leptanilla japonica*) even have "taps" on their body through which the workers sip their hemolymph.

∨ The workers of several species of ant are known to drink the hemolymph of their larval siblings (*Amblyopone* sp.).

CASTES

In the colonies of eusocial insects there is a division of labor, with individuals dedicated to collecting food, tending young, guarding the nest, laying eggs, and founding new nests. These workers, nursemaids, soldiers, queens, and kings often look wildly different.

Queens

In a eusocial insect nest the reproductive individuals are the queens, which can number one or a few per nest. The founding queen establishes the nest after a single, aerial mating (see Flying Ant Day page 139). These queens are generally long-lived and some rank as the most long-lived insects of all. A honey bee queen might survive for as long as four years, whereas a queen wood ant (*Formica* spp.) down in the deepest, most sheltered parts of her unmistakable nest might live for anywhere between 15 and 20 years. Queens of some other ant species might even live for 30 years. Termite queens can live for 20 years, possibly a great deal longer.

Life is not much fun for these queens. They might get tended to by the workers and have everything an insect could want, i.e. food, but they have to pump out an almost continuous stream of eggs. Any slacking in the egg-

∨ Termite queens are egg-laying machines. In their long life they can churn out millions of offspring.

production department will see the demise of the colony. In some termite species, the queen lays around 20,000 eggs every day from a bloated, sausage-like abdomen. There are also reproductive males, known as kings or drones. On the whole, all they have to do is provide sperm to a young queen, so she can found a new colony.

Workers

The other broad category of individuals in eusocial insects are the workers, which are responsible for everything else. Specialist workers called soldiers defend the colony, and in some cases, these are formidable enough to see off nearly all threats. For example, the soldiers of army ants and driver ants have enlarged heads packed with muscle to power their enormous mandibles.

In some of the leafcutter ants, there can be several, distinct types of worker. The smallest of these are known as minims and they take care of the brood and tend to the fungi gardens. Next up are the minors that continuously patrol and defend the foraging lines. It's normal to see a minor worker hitching a ride on a piece of leaf being carried back to the nest, not because they can't be bothered to walk, but because the worker carrying the piece of leaf is at risk from parasitoid flies that can develop inside the ants. Like a tiny guard dog, the minor worker keeps these flies at bay.

> Small leafcutter ant workers ride on leaf sections carried by larger workers to defend their larger sisters from parasitoid flies.

∨ To defend the colony, army ant soldiers have large heads packed with muscle to power the mandibles.

Medium workers are bigger than the minors and their main job is collecting foliage and returning it to the nest. A few years ago, it was discovered that these medium workers change tasks when the cutting edge of their zinc-reinforced mandibles, which start out as sharp as a razor, begins to dull and the task of cutting the leaves takes longer. These older medium workers then shift to carrying duties.

Largest of all the leafcutter workers are the majors. These are the colony's soldiers, but they also do heavy lifting, such as clearing the foraging trails of larger bits of debris and carrying bulky morsels back to the nest.

∨ The workers in social insect colonies shift huge quantities of material to build and maintain their nests.

Caste Development

The vastly different forms we see in an ant or termite colony are even more remarkable when we consider that they have the same DNA. How do you get adults that look, function, and behave so differently from just one set of instructions? In some, if not all cases, this all depends on what the young insect is fed. Giving lots of food, not enough food, or a specific type of food at the appropriate time appears to flick a developmental switch, which results in the staggeringly different adult forms.

^ Workers go to great lengths to tend their developing siblings. Here, a honey bee tends to a larva in its cell in the hive.

v The vastly different forms we see in a single colony of social insects—here, termites—are all derived from one set of genetic instructions.

When a honey bee nest is ready to produce a new crop of queens, each of which will hopefully go and start their own colony, the workers make queen cells and feed the larvae in these cells copious amounts of royal jelly, a moreish substance secreted from glands in their head. Feeding this stuff to the larvae in the queen cells throughout their development triggers the queen form, complete with fully developed ovaries for having their own offspring.

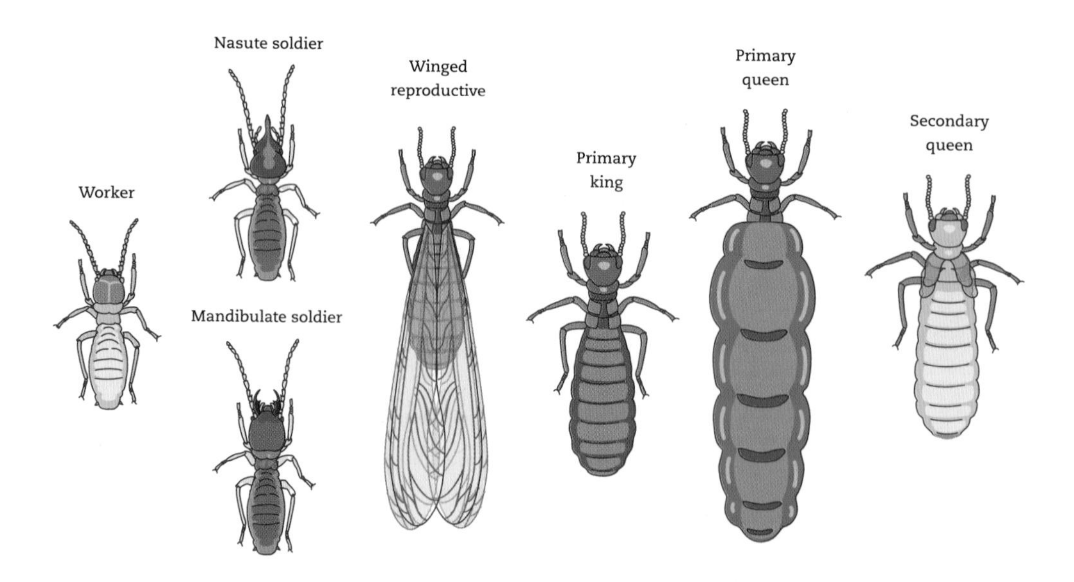

FLYING ANT DAY

If you're a fox or a bird you'll love flying ant day as the ground and the air will be alive with lots of edible morsels, but if you're scared by insects you'll hate it. Love it or hate it, the mass emergence of flying ants is a nice little peek into the lives of these otherwise secretive and mostly subterranean animals. Rather than just one day, the mass emergence and flights of these insects may be multiple events spread over a month, depending on temperature and humidity. These events can involve so many ants that they show up like clouds on weather radar. But what is going on?

The single goal of an insect colony is to produce more colonies and within the depths of the nest, new queens and males are nurtured that will hopefully succeed in doing this. What you're seeing during flying ant days are nuptial flights. Lots of would-be queens and male ants taking off from nests across the whole landscape. This will be the only opportunity for females and males from different colonies to find each other and mate—a very brief fling in the air—after which the male promptly dies and the female descends to the ground, gets rid of her wings and, if she's very fortunate, establishes a new nest. This requires a good dollop of luck and only a tiny proportion of them will be successful. Emerging *en masse* like this also swamps predators, improving the chances of any one young queen dodging all the predators and starting a new colony. There are also flying termite days, when huge numbers of winged female and male termites emerge to do the same thing.

∨ Emerging *en masse*, means that new queens can find a mate and swamp predators (R). The chances of any one of them founding a successful new colony are tiny. After the event the males die and fall to the ground (L).

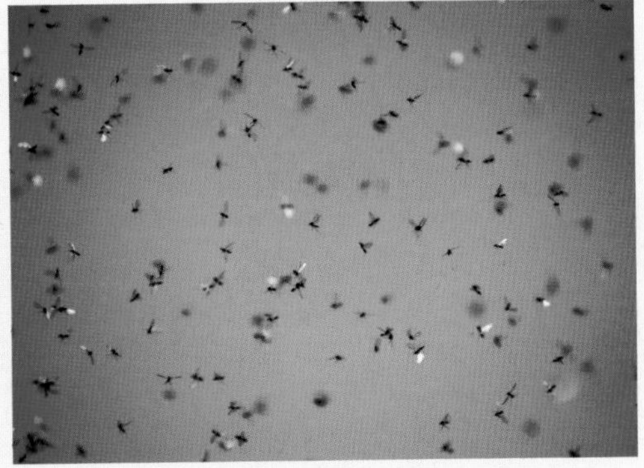

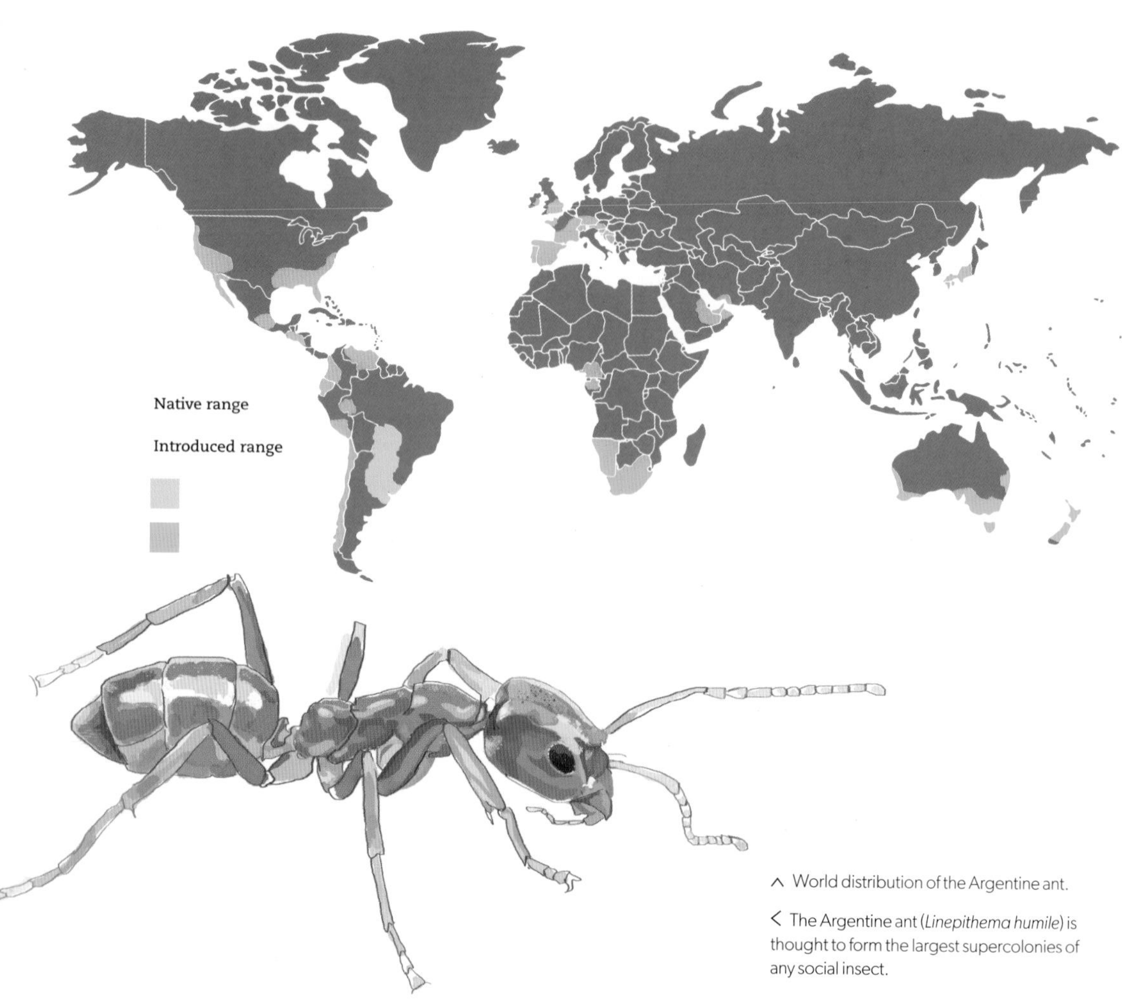

^ World distribution of the Argentine ant.

< The Argentine ant (*Linepithema humile*) is thought to form the largest supercolonies of any social insect.

Ant Supercolonies

The popular view of wasp, ant, and bee societies is a group of workers inhabiting a single nest ruled over by a single queen. Painstaking research involving genetics and the marking of thousands of worker ants has shown that the reality is much more complex. For example, studies show that wood ant colonies can have a single queen (monogyny) or lots of queens (polygyny); and the colony can inhabit a single nest mound (monodomy) or multiple, connected nest mounds (polydomy) that form by "budding."

In any given area, there might be a mix of wood ant social structures; from one colony occupying one nest with a single queen to supercolonies occupying many separate but connected nests with hundreds of queens.

At the most extreme end is a species of wood ant from Japan (*Formica yessensis*) that forms gigantic supercolonies, covering an area of 1.04 miles^2 (2.7 km^2) and containing an estimated 306 million workers and one million queens. Even these wood ant supercolonies might be dwarfed by those of the Argentine ant (*Linepithema humile*), which live in vast numbers across Europe, the US, and Japan. This seems to be a single mega-colony made up of billions of ants that has colonized a significant portion of the globe and is still growing. The individuals from different continents refuse to fight each other and act as though they're related, which gave biologists the first inkling that they were dealing with a mega-colony. In Europe, one enormous arm of this mega-colony is thought to stretch for about 3,730 miles (6,000 km) along the Mediterranean coast. In the US, another arm extends over 560 miles (900 km) along the coast of California. There is also a third huge arm extending along the west coast of Japan.

The budding strategy that forms these supercolonies and mega-colonies may be a relatively low-risk way of expanding into new areas, because the chances of a queen dispersing far from the colony and successfully establishing a new nest are very low.

Slave-making Ants

Founding a colony from scratch is fraught with difficulties and few queens are successful. Around fifty species of ants—with many more still to be described—have evolved a sinister means of setting up a colony. They enslave individuals from other ant species. The slave makers raid the nests of related ant species, steal the brood, or even confuse active workers into leaving with them. During these raids, they can sow seeds of confusion by secreting chemicals that make the host workers attack each other rather than the raiders. Back in the slave-maker's nest, the enslaved can still fight back, with some species systematically pulling the slave-maker pupae apart or carrying them outside to perish.

∨ Left: Plenty of ant species are slave makers. Here, a queen of the slave-maker species *Polyergus lucidus*, is shown with the workers and brood of the host, *Formica archboldi*.

Right: Slave-makers (*Polyergus lucidus*) returning from a raid on another of their host species (*Formica incerta*). Two of the latter already incorporated into the mixed colony are visible to the right of the nest entrance.

DEFENDING THE COLONY

The nests of eusocial insects are jam-packed with all sorts of goodies that other animals want. The colony's brood—an abundance of nutrient-packed eggs, larvae, and pupae—is a big draw. There are also food stores and piles of refuse, all of which are coveted by rafts of nest raiders. These nests are also safe, calm places to be. They offer protection and stable conditions.

The nest interlopers range from the benign, which do no obvious harm, to other more insidious species that do unpleasant things to the nest's occupants. To keep these undesirables at bay, eusocial insects have evolved all sorts of defenses from the thick walls of the nest itself through to chemical weapons and kamikaze soldiers.

∨ Social insect nests are packed with resources, so soldiers and fortifications are a key aspect of life for these animals.

Termites

Among the termites, the conspicuous mounds some species construct are their first line of defense. These mounds can persist for decades, even centuries and are made from a combination of soil and termite bodily secretions, which form a cement that is many times harder than the surrounding ground. As well as fortifications, these mounds are ingenious, climate-controlled high-rises. They harness the wind to ventilate the subterranean colony.

Any breach in these defenses gets the occupants excited and piling in to defend the colony. Obviously, it's the soldiers that step up to the plate first. Some termites employ a passive form of defense where they simply use their body to plug the breach in the nest. This is known as phragmosis. These phragmotic soldiers have a reinforced head, which coupled with their gnashing jaws is an effective way of protecting the breach until it can be repaired.

Beyond the pluggers, we see a devilish array of weapons in termite soldiers. There are jaws for piercing, jaws for slashing, jaws for crushing, and jaws that snap together to deliver fierce, lethal blows to enemies. Weirdly, some soldier termites have nearly lost their jaws altogether and are instead equipped with nozzles for squirting noxious chemicals over enemies or have a brush-like structure for daubing poison.

The chemicals deployed by soldier termites in the defense of the colony are varied and special. Depending on the species, the soldiers produce greases, irritants, contact poisons, and glues. Greases are used by soldiers with slashing jaws and they are known to slow the healing of the wounds caused by the jaws—in ants at least. Irritants force attackers, such as ants, to stop their assault as they try and frantically clean themselves. Contact poisons enter the enemy through wounds made by the jaws and encourage it off this mortal coil. The glues produced by soldiers are effective in entangling enemies and are often combined with irritants and poisons. All in all, termite soldiers have a real chemical armory at their disposal.

The most remarkable termite soldiers are the kamikazes. The soldiers of some species literally poo themselves to death. Over-enthusiastic defecation when defending the colony can force the soldier to burst, covering the enemies with the contents of various defensive glands. Tar-baby termites can rupture their body at will, entangling attackers in noxious, quick-setting gunk. These most drastic defenses are not limited to the soldiers either, the workers of these species can also defecate until they burst or turn into a puddle of sticky mess.

∨ Nasute soldier termites have a nozzle on their head for squirting noxious chemicals over their enemies.

∨ In the exploding ant, *Colobopsis saundersi*, a pair of over-sized mandibular glands run the entire length of the body. These can be ruptured at will, covering enemies in toxic gunk.

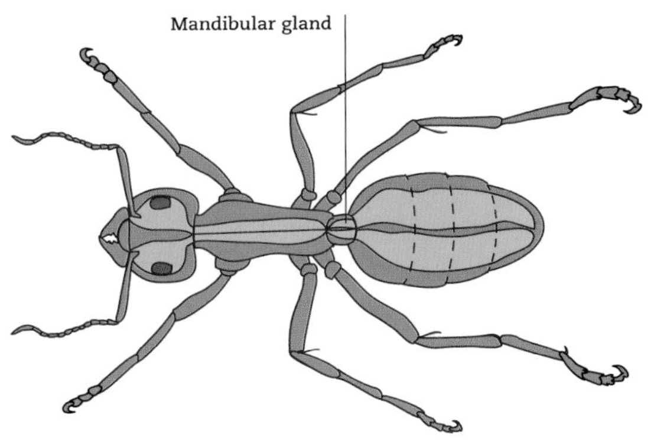

Ants

Ants are similar to termites in lots of ways, including their varied defenses. They too can build very strong nests that can resist all but the most specialist plunderers. There are also "plug" workers that use their body to breach holes in the nest. They too have evolved an extraordinary range of mandibles. Among the most impressive of these are those of the so-called trap-jaw ants. These fearsome gnashers can be "cocked" in a 180° gape and then released to close with an astonishing speed that flings intruders out of the way or catapults the worker out of harm's way. There are even kamikaze exploding ants that rupture their abdomen at will to release a bright, toxic gunk.

One thing that ants have that termites don't is a sting. Most ant species are capable of stinging and many of them can inject extremely potent venom. The Maricopa harvester ant packs the most potent insect venom known. Gram for gram, this ant's venom is much more lethal than that of many deadly snakes. Only tiny amounts of venom are injected when this animal stings, but in humans this is sufficient to cause intense pain that lasts for several hours. The most infamous, almost legendary, stinging ant though is the bullet ant. This is a big ant and its sting is bigger still; some say it is the most painful of any insect.

In some cases, ants smear or flick venom over their enemy. This is the tactic of the well-known fire ant (*Solenopsis* spp.), but when it squares up to the superbly named raspberry crazy ant (*Nylanderia fulva*) it comes unstuck as the crazy ant smears itself with its own venom, which neutralizes that of the fire ant. The stinger and its associated glands have been further modified in other ants to allow noxious liquids to be squirted out under pressure. Wood ant workers defend their nest by bending their abdomen through their legs and squirting formic acid at the threat, whether that be a bird, a badger, or a bespectacled naturalist.

Left: Phragmotic ant and termite workers plug nest entrances and breaches with their flattened, reinforced heads.

Right: Plenty of ant species have painful or very irritating stings they use to good effect against larger animals. The bullet ant has a very painful sting.

^ Unlike their smaller relatives, giant honey bees build their nests in the open, attached to branches or under rocky overhangs. As a result they've evolved a number of defenses to keep predators at bay.

Bees and Wasps

Of all the social insects, the ants and termites have the greatest range of defenses, but the bees and wasps also know how to look after themselves. These have all sorts of sophisticated security to counter their enemies, from hovering guards, guards at the nest entrance, and hot "bee balls" (see Japanese Honey Bee, page 146).

Equipped with a substantial sting and powerful mandibles, the giant honey bee of south and southeast Asia is an intimidating beast, but its enormous nests, rather than being built inside some sort of cavity, dangle free from the underside of branches or rocky overhangs. For this reason, this species is probably the most defensive of all the honey bees, even the African honey bees, which are well-known for their grouchiness.

First off, the comb of the nest brimming with food stores and the colony's brood is protected by a curtain of grumpy workers, several individuals thick. If the very sight of 100,000 or so enormous, angry honey bees dangling from a branch 100 feet (30 meters) or more up a tree is not enough to dissuade a hungry bear or bird, the bees will give it the insect equivalent of a Mexican wave. All the workers clinging to the exposed nest will lift their abdomens sequentially creating what looks like a shimmer. This lets birds, mammals, and predatory wasps know they've been spotted and that further action will be forthcoming if they persist. If the predator plods on regardless, a guard worker will fly in a zigzag pattern along the curtain of bees with its stinger poking out. This behavior prompts other workers to scuttle to the lower edge of the comb, cling together to form thin chains and emit a hissing sound. This makes the nest look bigger than it is, and if a predator tries to get a mouthful of comb from the bottom edge of the nest it will end up with a face full of angry worker bees. Remarkably, not even these safeguards are enough to put off the most determined predators. Animals, such as sun bears, routinely raid giant, honey bee nests and make off with the booty. This is just one more reason to respect honey bears.

JAPANESE HONEY BEE

In Japan, the hives of European honey bees (*Apis mellifera*) are raided by the giant Japanese hornet (*Vespa mandarinia*). This hornet is a formidable brute of an insect, which is in fact one of the largest living wasps. When a hornet locates a hive of European honey bees, it leaves a pheromone marker all around the nest, and before long its nest mates pick up the scent and converge on the beehive. The hornets fly into the beehive and begin a systematic massacre. The European honey bee is no match for the hornet as it is one-fifth of the size. A single hornet can kill forty European honey bees in one minute, and a group of thirty hornets can kill a whole hive, something in the order of 30,000 bees, in a little over three hours. The defenseless residents of the hive aren't just killed either, but horribly dismembered. The hive, after one of these attacks is littered with disembodied heads and limbs as the hornets carry the thoraxes of the bees back to their own nest to feed their ravenous larvae. Before they leave, they also gorge themselves on the bees' store of honey.

This amazing natural phenomenon begs the question: well what about the native Japanese honey bees? Do they get attacked? The answer is no and the reason is particularly neat. The hornet will approach the hive of the Japanese honey bee and attempt to leave a pheromone marker. The Japanese honey bees sense this and emerge from their hive in an angry cloud. The worker bees form a tight ball around the marauding hornet, which may contain 500 individuals. This defensive ball, with the hornet at its center, gets hot, aided not only by the bees vibrating their wing muscles, but also by a chemical that they produce. The hornet, unlike the bees, cannot tolerate the high temperature and before long it dies, and the location of the Japanese honey bees' nest dies with it.

> Top: The Japanese honey bee has a evolved a means of dealing with giant Japanese hornets—they cook them.

Bottom: Queen giant Japanese hornets (*Vespa mandarinia japonica*) are among the largest wasps. The workers of this species often raid honey bee nests.

GUESTS OF SOCIAL INSECTS

Any animal that manages to live inside a social insect's nest without being hounded and pulled apart like soft bread by the workers is on to a winner. These nests are fortresses and they offer a stable temperature and humidity. These desirable living conditions are topped off with an abundance of food, all of which has lured a great many "guests" to the nests of social insects.

It is in the nests of termites and ants where we find the greatest diversity of guests, but social bees and wasps also have their own fleet of freeloaders. These guests are as diverse as they are strange. There are snails, woodlice, millipedes, pseudoscorpions, schizomids, spiders, silverfish, cockroaches, crickets, butterflies, moths, flies and last, but not least, lots of mites and beetles. There are even parasitic social insects that live in the nests of other social insects. In many cases, the guests have hacked the code of the host's odor-based messaging system. By mimicking the odors of the host, they are often treated as nest mates, even nurtured by the workers. Probably the best studied of all these are the blue butterflies and their nefarious ways (see Blue Butterflies, page 153–154).

< The tank-like larva of a *Microdon* hoverfly. These wander the nests of their ant hosts. Depending on the species these larvae are active predators or scavengers (*Microdon testaceus*).

Masters of Exploitation

As a group, beetles have really gone to town exploiting the nests of social insects and few other guests can match them for sheer strangeness. Take the very weird rove beetles that live in termite nests and have come to mimic their hosts, complete with an enormously swollen abdomen. Some also have sausage-like appendages dangling from their distended abdomens that look a bit like legs when viewed from above—adding to the termite illusion. The termites have also been observed licking these danglers, perhaps being duped by their chemical secretions into treating the beetle as a fellow termite.

< Top: Lots of flies have taken to living in social insect nests. This unusual phorid fly lives with the fungus-growing termite *Odontotermes* (*Javanoxenia* sp.). Many of these termite guests have large bloated abdomens (physogastric), which give them a superficial resemblance to their hosts.

Bottom: The woodlouse, *Platyarthrus hoffmannseggi*, lives in the nests of various ant species.

∨ The tiny scarab beetle *Termitotrox icarus* being carried by its termite host (*Odontotermes proformosanus*). This beetle was only described in 2020 after being discovered in the fungus gardens of its host.

Other beetles have become masters at exploiting the nests of ants. These range from species that wander in or near the nests to prey on the ants, to species that are completely integrated into the nest, and are fed and cleaned by the tireless workers. To add insult to injury, the latter normally feed on the host's brood and have all sorts of adaptations to grease the wheels of acceptance, including golden tufts of scales, fancy antennae that secrete host-mimicking odors, and chemicals that appease or startle the workers when they become suspicious.

Paussine bombardier beetles are one such group of specialist nest predators. Like malevolent house guests, they wander through the galleries and chambers of the nest, smelling of ant, helping themselves to their host's brood. Occasionally, a worker ant will grow suspicious as the greedy beetle demolishes yet another plump larva, but under ant interrogation the beetle simply exudes more ant aroma from its bulbous antennae and the pores on its body, and the suspicions of the worker are quickly allayed.

> Top: These unidentified silverfish live in termite nests.

Bottom: The smallest cricket species live in ant nests. These are well integrated with their hosts and can often be seen receiving droplets of food from the host workers (*Myrmecophilus* sp.).

∨ Rove beetles are the most diverse interlopers by quite some margin in the nests of social insects. This *Longipedisymbia* species lives with the termite *Longipeditermes longipes*.

HORNET'S NEST
(Vespa spp.)

Who'd want to live in a hornet's nest? Hornets are big, powerful wasps with quite a reputation, yet even their nests are used and abused by other animals. If you're brave and you want to root around a European hornet (*Vespa crabro*) nest you will find the larvae and adults of the splendid rove beetle *Quedius dilatatus*. Exactly what this beetle gets up to is unknown. They probably hunt fly larvae in the mound of detritus that forms beneath the hornet's nest, but they may also eat dead and dying hornets that fall from the nest. Lots of flies live in these nests as larvae, including the impressive hornet mimic, *Volucella inanis*, the larvae of which eat the host's larvae.

> The larvae of the hoverfly, *Volucella inanis*, develop in the nests of several social wasp species, including the hornet, *Vespa crabro*.

∨ The larvae of this rove beetle develop in the detritus beneath hornet nests, feeding on waste and possibly on dead and dying hornets (*Quedius dilatatus*).

BLUE BUTTERFLIES

Butterflies are capable of behaving in some very despicable ways, especially as caterpillars. The adults are fleeting, shallow, lustful creatures, only interested in getting it on with the opposite sex, but the larvae have the difficult task of eating as much as possible in the shortest amount of time. To do this, some species have found that they can get the job done very successfully if they pull the wool over the beady eyes of ants.

Blue butterflies (*Phengaris* spp.) lay their eggs on the flowers of their food plants. When a caterpillar hatches it feeds on the flowers for a couple of weeks, but a change eventually comes over it. During the early morning or evening, the caterpillar chews its way from the base of the flower and shuffles along the petals to the apex of the bloom. Apparently tired of its nursery, it releases its grip and falls to the ground on a silken thread and waits.

˅ Many butterfly species across two families are closely associated with ants. Some of these butterflies are among the most well studied guests.

This is the riskiest time of its short life. Predators abound among the short turf and all of them would make short work of a tiny, plump caterpillar; however, a small, foraging red ant (*Myrmica* spp.) gets a whiff of the caterpillar and goes for a closer look. The ant, seemingly intrigued and mesmerized by the caterpillar, strokes it all over with its quivering antennae. If it could, the caterpillar would be breathing a huge sigh of relief as this is exactly what it was waiting for. This is a result. To express its relief, the caterpillar produces a drop of sweet fluid from its rear end, which the ant immediately starts suckling. This can go on for some time, until the caterpillar flattens the middle or rear of its body, a simple act that is apparently enough to completely fool the ant that the caterpillar is a grub from its own nest that has somehow gone walkabout. It tenderly picks the caterpillar up in its jaws and makes for its nest.

The caterpillar is deposited in the nursery of the ant's nest alongside the countless young of the colony. Here, it blends right in. It smells right and smell to ants is all important. The ants feed the caterpillar by regurgitating nutritious fluid, and some of these tiny tricksters persuade the ants to give them preferential treatment so that they receive more attention and food than the ant grubs. They do this by producing low-frequency sounds that mimic those sounds made by the queen. This lavish care even extends to the parasitic caterpillars being rescued before the ant's own brood when the nest is disturbed!

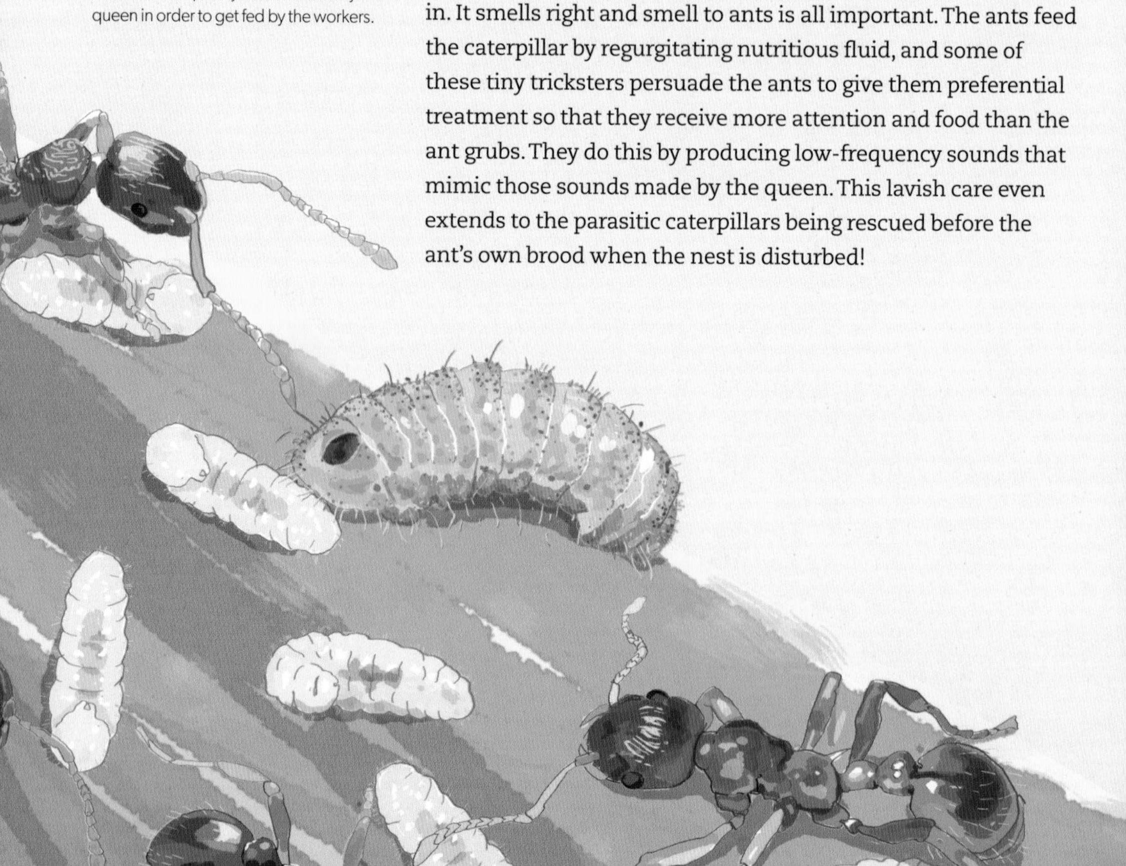

∨ Host worker ants will carefully tend the caterpillars. In some species, the caterpillars mimic low-frequency sounds made by the queen in order to get fed by the workers.

The caterpillars of large blue butterflies (*Phengaris arion*) do not get this same level of care, instead they feed exclusively on the ant brood. On such nutritious diets the caterpillar grows quickly, increasing its weight by about 100 times during the first month in the nest. The caterpillar remains in the nest for some time, living it up in the safety of the nest at the expense of the trusting ants and only when the summer arrives does it begin the transformation that will turn it into a fine butterfly.

When one of these blue butterflies emerges from its chrysalis deep underground as an adult butterfly, the tricks used to deceive the ants have long since worn away. It no longer smells like an ant and it certainly doesn't look like an ant larva. As the penny drops among the long-suffering ants, it's time for the butterfly to leg it. The ants see the butterfly they have nurtured for the best part of a year as nothing more than an intruder, and if they get hold of the low-down mimic they'll tear it apart. Fortunately, the butterfly has one last trick to avoid being held to account and as the angry ants try and bite the fleeing charlatan all they come away with is a mouthful of scales. The whole body of the newly emerged butterfly is densely covered in loose scales and the ants cannot get a grip on it, so by the scales on its behind it manages to give its trusting guardians the slip and leaves the nest by the nearest exit. Above ground, it makes for a perch among the lofty vegetation of the meadow where its wings will inflate and harden in the summer sun until it is ready to flutter off.

This is not the end to this remarkable story. It has recently been discovered that when these female butterflies are laying their eggs they are drawn to food plants that are close to the nests of their ant hosts, attracted to a chemical that is given off when the plant's roots are disturbed by nesting ants. Perhaps this is an attempt by the plant to suppress the ants by attracting their parasites? Not only that, but as with all struggles in nature, the ants and the butterflies are locked in an evolutionary race. The caterpillars dupe the ants by smelling like them, so populations of the host ants alter their odor signature over generations to catch the parasites out.

BUTTERFLIES TAKING ADVANTAGE

- More than 50 percent of lycaenid butterfly species (around 3,000 species) interact with ants during part of their life cycle
- Many metalmark butterflies (*Riodinidae*) are also associated with ants
- *Spindasis lohita* caterpillars produce three types of vibrational calls, which trigger certain ant behaviors

5

PARASITOIDS AND PARASITES

Parasitoids are the most fascinating insects. They are predators that consume a single prey item during the course of their development, feeding on it, or in it, and typically killing it in the process. We can find parasitoids across the insect tree of life in mantisflies, beetles, flies, wasps, moths, and caddisflies. The most diverse groups, by a considerable margin, are the parasitoid wasps and tachinid flies. Within some of these groups, this way of life has evolved independently many times, perhaps more than 100 times in flies. These insects, especially the wasps, are probably the most diverse animals of all.

A SUPER-SUCCESSFUL LIFE STYLE

In terms of species, parasitoids are the most successful insects and among the most striking when you look at them closely and understand a little about how they live. This is summed up beautifully by the entomologist A. A. Girault (1884–1941), when musing on parasitoid wasps: "Some gem-like or marvelous inhabitants of the woodlands heretofore unknown and by most never seen or dreamt of."

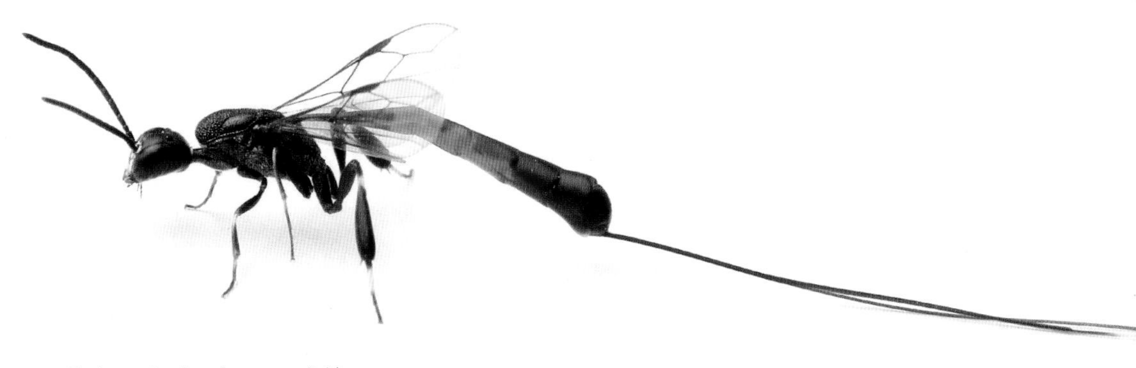

∧ The long ovipositor of some parasitoid wasps enables them to get to concealed hosts.

< Larvae of a parasitoid wasp (*Hercus fontinalis*) feeding on their host—a moth caterpillar.

Many of the bugs you're aware of, and those that you aren't, have at least one parasitoid, sometimes several. The individual stages in the life cycle of an insect—the egg, larva, nymph, pupa, and adult—can be attacked by one or more parasitoid species. Not only that, but the parasitoids often have their own parasitoids—the excellent hyperparasitoids (see Trigonalidae, pages 170–171).

For the most part, the lives of parasitoids are unseen and unknown. You might catch a glimpse of a large ichneumonid wasp scampering across a leaf, its antennae twitching feverishly, or a flaccid caterpillar studded with tiny silken cocoons spun by mature parasitoid wasp larvae as they emerge from their host, but that's about it. They're often fiendishly difficult to identify and many are incredibly tiny—among the smallest animals—so they are very easy to miss. In terrestrial habitats they're everywhere, and you're probably never more than a couple of feet away from a parasitoid.

Sweeping a net through a meadow for a few minutes will capture thousands of small animals. Dumping the contents of the net into a white tray will reveal a staggering diversity of living things. Some of what you see are large insects, such as bees and large flies, but beyond these, among all the seed heads and dry stems, is a constellation of smaller animals, many of which are on the very cusp of what can be seen with the naked eye—scurrying specks and ascending motes of flying dust. Many of these specks are parasitoid wasps. The small size of many parasitoids has enabled them to exploit all sorts of niches that are inaccessible to larger animals, but it has also presented them with some tricky challenges, especially reproduction as they're limited in the number of eggs they can produce (see Multi-wasp, pages 172–173).

∧ This parasitoid wasp (*Eupelmidae* sp.) is drilling down to a concealed host using its egg-laying tube (ovipositor).

∨ A huge proportion of insect eggs will fall prey to specialist egg parasitoids. This whole batch of shieldbug eggs was lost to a parasitoid wasp, a *Telenomus* sp.

Most parasitoid insects are either wasps or flies, but some of the others deserve a mention, too. Take the small Brazilian moth (*Sthenauge parasiticus*) that eats the venomous spines of other caterpillars. The caterpillar of this species spins a silken tunnel across the body of its host from where it feeds on the adjacent, venom-loaded spines. This is not ideal for the host, and it gradually weakens and dies. The parasitoid caterpillar is not averse to feeding on the dead host, and will bore into it to feed on its insides. A strange group of moths make their living sucking planthoppers and cicadas dry. The young caterpillar attaches itself to the hopper/cicada, plunges its mouthparts into the host and drains the life out of it over a period of four to five months.

Parasitoid Beetles

Beetles have joined the parasitoid party too, although this is not a common way of life among these insects. A group of beetles with elaborate antennae are parasitoids of cicadas. The active first instar larva searches out the host—a young cicada nymph—and latches on to it as it burrows into the soil to live a subterranean existence. When the host is big enough it starts eating with gusto. Some ground beetles have turned parasitoid, seeking out and feeding on the pupae of leaf beetles, perhaps even stealing the noxious chemicals of these well-defended hosts. The active, hatchling larvae of most oil beetles are parasitoids, hitching a ride on a bee back to its nest where it breezes into a brood cell, consumes the host's egg, and then eats the food stores, usually pollen (see Chapter 1 for more on these).

∨ Not many beetles are parasitoids. A notable exception are the Rhipiceridae, which are parasitoids of cicadas.

Complex Lifestyles

The host range and life styles of parasitoid flies is something else—just consider the ant decapitators (see overleaf). Beyond other insects and arthropods, they attack an extraordinary range of hosts. There are flies that develop inside terrestrial flatworms, freshwater snails, earthworms, and last but not least, the faces of frogs and toads. As larvae, conopid flies develop inside bees and wasps. The tip of the female's abdomen is often a bit like a can opener, which it uses to pry open the abdominal plates of the host to insert an egg. Only a few of these species have been studied in any detail, but it turns out the larval fly can somehow control the behavior of the host. In one species, the bumblebee host digs its own grave by burying itself in the soil just before it completely succumbs to the parasitoid larva munching on its insides. How the parasitoid larva does this is unknown, but it hugely improves the chances of the parasitoid pupating and emerging as an adult to bring more misery to unsuspecting bumblebees.

Host manipulation is probably common in parasitoids, but for the most part we have only scratched the surface of understanding the complex interactions that take place between parasitoids and their hosts. What with their comically tiny heads, the small-headed flies are rather innocent looking beasts, but these too are parasitoids—every species completes its larval development inside spiders. The female fly deposits a load of eggs on or near a spider's web, and the active larvae that hatch seek out and enter the host, often through a leg joint. From there, they wriggle to the arachnid's book lung and sit tight, sometimes for years, waiting for the doomed host to grow to the right size. When the time is right, the fly larvae somehow makes the spider spin a protective web and then consumes its host. Many parasitoid wasps do something similar, although they often sit on the outside of the unfortunate spider, draining the life out of it and forcing it to spin a special protective web before it inevitably succumbs. You might have

∨ A huge variety of parasitoid wasp larvae feed externally on spiders. Some even control the host's behavior (*Polysphincta* sp.).

∨ This tarantula has a young parasitoid wasp larva attached to its abdomen.

∧ The braconid wasp *Dinocampus coccinellae*, is a parasitoid of ladybirds. The zombified host stands guard over the parasitoid's cocoon.

even seen a parasitoid wasp larva that controls the behavior of its host—a ladybird. The mature wasp grub emerges from the beetle and spins a cocoon beneath it. The beetle stays astride the cocoon—zombified and occasionally twitching—giving the pupating wasp some defense from predators. Amazingly, the beetle can sometimes survive this.

> Small-headed flies are specialist spider parasitoids too. Their life cycle is long, convoluted, and fascinating.

ANT-DECAPITATING FLIES

In many places worker ants are on high-alert for the danger hovering silently above their heads—female ant-decapitating flies who are ready to lay eggs. Spying a suitable target, a fly makes a darting flight downward and lands delicately on the back of the ant, which is sometimes many times her size. She is very picky about the ant species she selects. Some ant-decapitating flies will only prey on one species of ant, whereas others may use a handful of species. The fly probes around the ant using her sharp egg-laying tube and with the dexterity of a seamstress she pierces the thin membrane between the plates of the ant's exoskeleton. Depending on the fly species in question, the egg is laid in the head, thorax, or abdomen. With a single egg deposited, the fly takes off to search for more potential hosts.

In the meantime, a larva eventually hatches from the egg and if it was deposited in the abdomen or thorax it will wriggle and squirm its way up to the ant's head. Once in the snug head capsule, the larva settles down to feed, gorging itself on the muscles and other tissues that fill the head of the hapless host. The development of the larva can be very rapid and soon enough the head has been emptied of all edible matter. The head capsule drops off and the fly larva completes the rest of its development

∨ A tiny ant-decapitating fly (*Microselia* sp.) perched on the abdomen of its host *Camponotus japonicus*.

in the safety of this small shell. In an animal like a mammal, where all the really important bits of the central nervous system are seated in the brain, such a fate would undoubtedly end in death. However, the nervous system of an ant has several small brains along its length. These ganglia can control walking and other activities, and as long as it has sufficient food reserves it may soldier on for a while until it keels over. In other cases, the head capsule may remain attached to the ant, but it is completely empty save for a well-developed grub or pupa of the ant-decapitating fly.

∨ Top left: A female phorid fly, *Pseudacteon* sp., hovers over a fire ant worker.

Top right: An ant that has been decapitated as a result of *Pseudacteon* parasitism.

Bottom left: Two pupae of a *Pseudacteon* fly (dorsal and ventral views) next to a pupa still in its host's head. Note correlation of size and shape to ant host, hardened and darkened anterior portion, and respiratory horns.

Bottom right: An adult *Pseudacteon litoralis* emerges from an ant head.

Tachinid Flies

The most extraordinary parasitoid flies are the tachinids. This is a big group of insects, with around 10,000 described species and probably many times more that number still to be described. Like many of the parasitoid wasps, these flies are everywhere and I guarantee you've seen lots of them. They seek out and develop inside a huge range of other arthropods, but caterpillars, sawfly larvae, beetles, and bugs are the most frequent victims. Most of these flies lay eggs that hatch almost immediately, and the female flies often inject the eggs straight into the host. This strategy comes with its own set of challenges, not least the immune system of the host. Yes, insects might be small, but they have a sophisticated, innate immune system for dealing with pathogens, parasites, and parasitoids. Once in the host's body cavity, the parasitoid eggs are at the mercy of the immune system. Special blood cells can surround and encapsulate the parasitoid's eggs, ultimately killing them. Tachinids can sidestep these defenses by somehow distorting the host's normal wound-healing process and immune response, which ultimately forms a funnel through which they can breath and simultaneously eat the host's insides.

Finding a Host and Surviving Inside It

One of the biggest challenges for parasitoids is finding their hosts. They do this by detecting odors given off by the host, its waste, or the food it's eating. When under attack from herbivorous insects, lots of plants release chemicals that parasitoids use to locate their hosts—effectively these signals are a plant-style SOS. Hosts may be deeply concealed in solid wood, so parasitoids wasps have a form of echolocation. Using antennae, which are often thickened or hardened for the purpose, the wasp taps the wood and then listens for the echoes using special organs in its legs. Using this technique, the wasp can pinpoint the host with devastating precision. Reaching these hidden hosts is another matter, but the wasps have just the tool, an egg-laying tube (ovipositor) that is modified into a drill reinforced with tiny quantities of metal ions, such as zinc, manganese, or copper. The wasp drills down to the oblivious

∨ Tachinid flies are among the most diverse and abundant insects, yet are easily overlooked. This species, *Dexiosoma caninum*, is a parasitoid of beetle larvae.

∨ Tachinid eggs (possibly *Trichopoda pennipes*) on the western conifer seed bug (*Leptoglossus occidentalis*).

host, delivers a paralyzing sting and then lays an egg on it or nearby. In some cases, the venom injected by a female parasitoid wasp does bizarre things to the host (see Zombie-makers, page 168–169).

Rather than simply laying an egg near an incapacitated host, many parasitoid wasps inject their eggs directly into the host, again using the ovipositor. In some circumstances, the dastardly wasps can defeat the host's immune system with the help of proteins, virus-like particles, or even viruses. That's right, some of these parasitoid wasps use biological warfare (see pages 191–192).

> There are even parasitoid moths. The white caterpillars of this species (*Epipomponia nawai*) are feeding on a female cicada (*Tanna japonensis*).

∨ This wasp is a specialist egg parasitoid. The female wasp lays her eggs in the eggs of scale insects (*Wallaceaphytis kikiae*).

ZOMBIE-MAKERS

Some parasitoid wasps use their venom to control the behavior of their host. The emerald cockroach wasp is one such insect. This flying jewel preys exclusively on cockroaches. Using its powerful senses it homes in on an unwary victim and administers two stings. When delivering a sting, the wasp faces the cockroach and curves its flexible abdomen around to inject the venom.

The first of these stings is directed at a tiny node of the central nervous system located in the insect's thorax. This mini-brain controls the cockroach's front legs and the wasp's venom blocks its activity, paralyzing the victim. This paralysis is only temporary, lasting for between two and five minutes. This is more than enough time for the wasp to deliver its second sting, which requires even greater surgical precision. Using its sensitive sting, it delivers a tiny dose of venom to a region of the cockroach's brain, which affects, among other things, its escape reflex. When it eventually recovers from the paralysis of the first sting, it doesn't try and flee for the nearest cover, but grooms itself excessively for around thirty minutes, while the wasp scuttles off to look for a suitable lair. When the wasp returns, it bites off one of the cockroach's antennae and laps at the blood that flows from

∨ The venom of the emerald cockroach wasp changes the behavior of the host, turning it into a docile, self-cleaning banquet for a wasp grub (*Ampulex compressa*).

PRECISION STINGS

- A sting is a modified egg-laying tube (ovipositor)
- Sensory structures on the sting of a hunting wasp enable it to pinpoint the exact target within the prey
- Very few solitary wasp species have been studied in any detail

∧ The emerald cockroach wasp seeks out cockroaches as food for its offspring (*Ampulex compressa*).

∨ Parasitoids, such as *Ampulex compressa*, are superbly adapted to find and subdue their hosts.

the severed appendage. Rather than just a welcome snack, this might also allow the wasp to check the dosing of its venom. The wasp then conveys the cockroach to the refuge it found earlier, leading it by the remaining antenna like a docile pet. There, the wasp lays a single egg on the zombified host.

The cockroach, essentially incapacitated, but still alive, is sealed in this hideaway with small stones and other debris, not to prevent it from escaping—it has no urge to—but to keep it safe from predators. The wasp larva hatches to find itself sitting on a mound of self-cleaning food, which it starts tucking in to. After two days, the young larva is big enough to tunnel into the host, and after four or five days the voracious feeding of the larva takes its toll and the cockroach dies. After about eight days, the wasp larva is ready to pupate and it spins itself a silken cocoon inside the drying casket of the cockroach. The adult hatches after about four weeks and leaves the lifeless husk of its host.

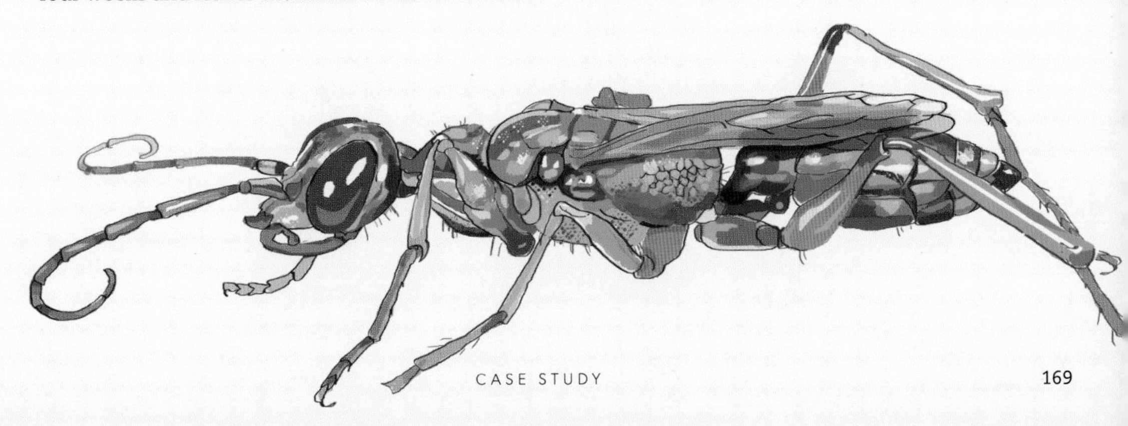

TRIGONALID WASPS

Among the huge diversity of parasitoids there are some bewildering life cycles, but there are few that can compete with the trigonalid wasps for the peculiar way in which they reproduce.

In most cases, a female parasitoid wasp deposits her eggs on or in the host, but this is far too pedestrian and safe for the trigonalids. These mavericks of the wasp world like to make things more difficult for themselves, so the female deposits her eggs on plant leaves. If this were the end of the story these wasps would not have lasted very long. No, the egg-on-the-leaf-trick is merely a ruse. The female trigonalid uses her short ovipositor and the unique structure of her abdomen to brace the leaf as she lays her eggs. She deposits a few eggs on one leaf before moving onto another leaf perhaps on a nearby plant. This goes on until she gets eaten or exhausts her eggs, which in some species of trigonalid can be as many as 10,000, a huge number for a parasitoid, which underlines just how haphazard this strategy is.

The whole purpose of going through the laborious act of laying thousands of eggs in leaves is so that her offspring can get swallowed by a caterpillar or sawfly larva. If the very tough egg is lucky enough to be swallowed, it hatches, which is thought to be triggered by the physical action of being chewed and/or salivary secretions. For most parasitoids, getting into a plump caterpillar would be mission completed and cause for celebration, but no such luck for the tiny trigonalid larva. The tiny larva winds up in the caterpillar's gut and wastes no time in breaking out of there to gain entry to the host's body cavity. The trigonalid makes its way through the body cavity searching for its real quarry—other parasitoid larvae that are already in residence inside the host. Lots of parasitoids, such as other wasps and flies, are dependent on caterpillars and sawfly larva, and it is these the

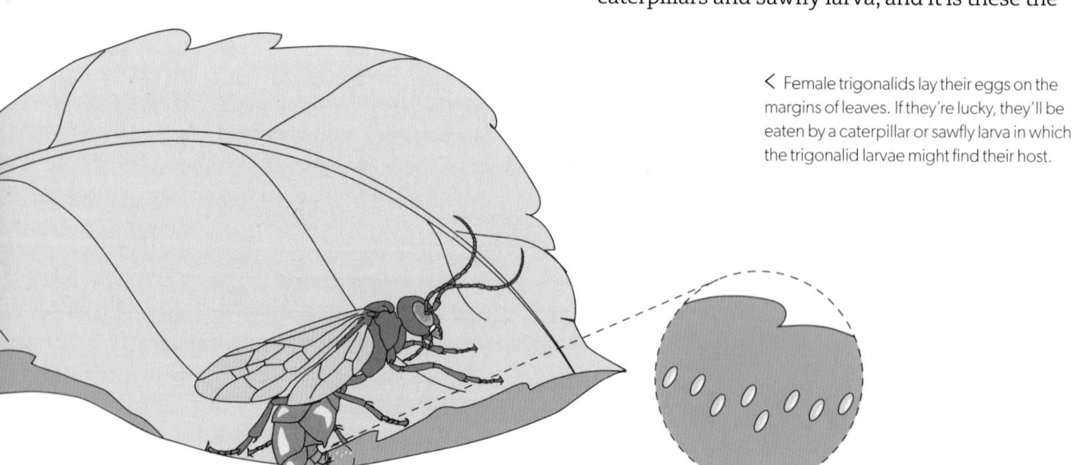

< Female trigonalids lay their eggs on the margins of leaves. If they're lucky, they'll be eaten by a caterpillar or sawfly larva in which the trigonalid larvae might find their host.

^ Trigonalids are rarely encountered. They have among the most complicated life cycle of any parasitoid. This is an unidentified species found in Peru.

trigonalid larva is after. If it's lucky, it will find its prey, attack it, and eat it; but in many cases the trigonalid that has defied the odds to get swallowed by a caterpillar will find nothing suitable to predate once inside, or the only parasitoid within the caterpillar will be too large for the trigonalid to tackle. In both cases, the poor little trigonalid is doomed. However, some species of trigonalid are able to sit tight inside the caterpillar until it does get parasitized by an ichneumon wasp or a tachinid fly.

This isn't the only bizarre life cycle of the trigonalids. Another strategy used by some species also hinges heavily on coincidence, but the supporting cast is slightly different. These species still depend on a caterpillar or sawfly larva, but this time the caterpillar must be captured by a wasp, butchered and fed to one of the wasp's grubs back at the nest. In the flesh of the dead caterpillar are the eggs of the trigonalid and once swallowed they hatch to feed on the unfortunate wasp larva.

In the world of wasps, trigonalids are something of an enigma. Only around 100 species are known, the adults don't live very long and their precarious way of life means they are rather rare (I've seen a live one just once), so we don't know a great deal about them. Their geographic distribution, appearance, and life style suggest they are very ancient. They may be something of a missing link between the sawflies and the rest of the Hymenoptera. Indeed, the oldest trigonalids are known from 100-million-year-old lumps of Cretaceous amber, which shows this way of life, precarious as it may be, has been going on for some time.

MULTI-WASP

Exactly how these tiny parasitoid wasps go about taking over their host is extraordinary. Many of them are the nemesis of certain moths and when the female wasp finds a suitable host egg she lays a single egg inside. In the most extreme cases, the egg, rather than just hatching into a single larva, goes through a remarkable sequence that sees this single egg become different types of larvae and eventually thousands of new wasps.

ATTACK OF THE CLONES

- A single egg of some tiny wasps can develop into thousands of identical embryos
- Soldier larvae of these wasps do not develop into new adults
- This type of reproduction is known in four families of wasps (*Platygasteridae*, *Braconidae*, *Dryinidae*, and *Encyrtidae*)

In the species that have been studied, the action begins in the host egg, but continues as the host larva develops. The host finally succumbs after it has spun a cocoon. Many other parasitoid wasps (for example, chalcidoids and *Platygastridae*) complete their entire development within the host egg and are also known to develop in this way—where one egg gives rise to lots of embryos (polyembryony).

The single wasp egg divides again and again to form a mass of undifferentiated cells, called a polygerm, suspended in the developing host embryo. Eventually this polygerm splits into discrete clusters of cells, each of which will become a wasp larva. Most of these wasp larvae are "normal" and will go on to ravage the host tissues, pupate, and escape from the remnants of their food as new wasps. A small proportion—about 10 percent in the species that have been studied—are special and develop faster and differently than the normal clones. These special larvae are anatomically simple, but are equipped with fearsome jaws. They are soldiers and they wander through the host to dispatch the eggs and larvae of other parasitoids that have laid eggs in the same host egg, even other clones of their species. Crucially, these soldier larvae don't pupate and will never give rise to new wasps. They are completely disposable and most of them will be dead by the time the normal larvae gorge on the host tissues.

The fate of each cluster of polygerm cells depends on whether or not they have germ cells (the beginnings of ovaries or testes). Those with germ cells become normal larvae and eventually wasps. Those without become dead-end soldiers, whose sole task is to rid the habitat of competitors. Remember, though, that all these larvae are clones, so it's not as though one individual is sacrificing itself for a genetically different individual. In some cases, this type of polyembryonic development can give rise to 3,000 wasp larvae from a single egg. These parasitoids are tiny and limited in the number of eggs they can produce, so this strategy allows them to get around this limitation. It also allows these species to adjust their brood size to the available resources.

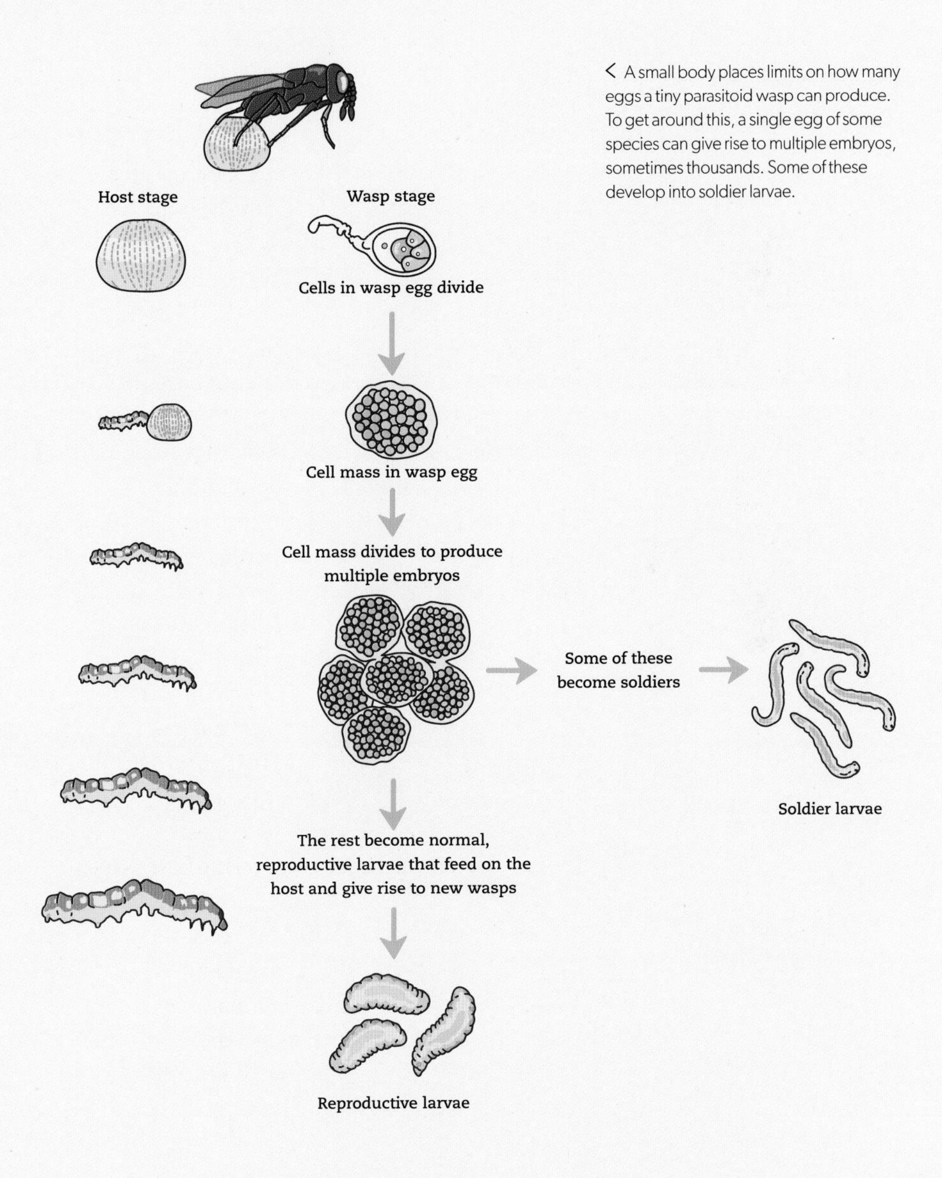

< A small body places limits on how many eggs a tiny parasitoid wasp can produce. To get around this, a single egg of some species can give rise to multiple embryos, sometimes thousands. Some of these develop into soldier larvae.

PARASITES

Compared with parasitoids, not all that many insects are truly parasitic. On the whole, we don't have much love for these animals because of what they do to us, our pets, and our livestock. There are the biting flies, chewing lice, sucking lice, kissing bugs, the itch-inducing bed bugs, the fleas, beetles that live on beavers, and moths that feed on tears and even blood.

In contrast to parasitoids, parasites normally feed on their host without damaging it so much that it dies. Naturally, there are exceptions to this—there always are. Parasitism has also evolved in lots of groups of insects, but it's among the flies and true bugs that we find the greatest variety of itch-inducing beasts. Many of these, such as the mosquitoes, are among the best-studied insects, primarily because we want to get rid of them.

Indeed, it is the flies that are the champion parasitic insects. Many of them have become specialist blood feeders and quite a few feed on the internal tissues of their host. The blood feeders, in particular, transmit a huge range of

∨ Some biting midges have become almost tick-like. They alight on a host—in this case a stick insect—and suck its blood until they become enormously bloated. They eventually drop off and lay eggs (*Forcipomyia* sp.).

< Mosquitoes are exquisitely adapted for exploiting the blood of larger animals. Only females feed on blood—they need it to mature their eggs.

diseases to large animals, just think of mosquitoes, but I don't want to dwell on that here as a library-worth of books has been written on this subject.

From a purely biological point of view, these parasites are fascinating animals. A mosquito can detect a host from a long way off, delicately probe the skin for a capillary, inject substances to kill any pain and dilate the blood vessels, before drinking its fill of the rich food source. The same can't be said for insects such as horse flies and deer flies, who simply slash open the skin with their sharp mouthparts. If you've ever been bitten by one of these animals you'll know that it hurts. Throughout the blood-feeding insects we can see evolution in action, from moths that have only recently turned to this way of life (see Clumsy Suckers, page 180–181), to old hands such as the mosquitoes.

In these blood feeders, it's the norm for only the female to feed on blood as she needs the protein to mature her eggs. The males often feed on nectar, which is much less risky. This is an important point. Stealing blood from a much bigger animal is fraught with danger. The bloodsucker could be swatted before it lands or splatted while its feeding, which is why many of these insects are small or inconspicuous and nocturnal, and why many of them have adaptations for extracting what they want as painlessly as possible. This food source might be rich in protein, but it's deficient in lots of other nutrients, so the guts of these blood feeders are brimming with microorganisms that can synthesize the missing nutrients and pass them on to the insect in exchange for food and a place to live. Another problem with this food source is that it's mostly water, and plugging your straw-like mouthparts into a blood vessel is a bit like drinking from a fire hydrant. These animals have to get rid of the excess water fast, otherwise they'll severely mess up their internal chemistry or burst like an overfilled water balloon. They have extremely efficient organs to do this—the insect equivalent of kidneys. Sap-sucking insects are faced with the same problem and this is why they secrete copious amounts of honeydew.

Some flies, such as louse flies, bat flies, and fleas, spend nearly all their life clambering around in the fur or feathers of their hosts,

sucking blood at will and trying their best to avoid the scratching and grooming efforts of their irritated hosts. Wings would be an impediment in this situation, so fleas and some of these weird flies have reduced or non-existent wings and enormous claws for keeping a good hold on their host. The fleas have become prodigious jumpers, using muscles and an elastic protein to propel them onto new hosts.

A few parasitic insects feed inside their hosts, and to our eyes at least, are probably some of the most gruesome insects. Bot flies feed beneath the skin of mammals and birds. Plenty of people who have been to Central and South America will tell you of their experience with the human bot fly. The warble flies live in a similar way inside cattle, deer, and sometimes an unfortunate human, but instead of feeding

∧ Deer flies, like this one from Amazonia and horse flies are specialist blood suckers too, although they simply slice the skin open to get at the blood (*Lepiselaga crassipes*).

∨ Insects are also plagued by parasites, many of which modify the behavior of their host in remarkable ways. Fungal fruiting bodies burst from the body of this Amazonian grasshopper (*Beauveria locustiphila*).

in one place, they go on a magical mystery tour through their host, tunneling through the tissues to the front end of the animal and then back to the lumbar region where they finish their development.

The stomach bot flies are even more peculiar. The female horse bot fly lays her eggs on the horse's legs, so they can be licked up by the host. From there, the larvae hatch, tunnel into the tongue and down into the stomach where they grip tenaciously to the stomach wall with powerful mouth hooks, scraping the surface to feed on blood. The fully grown larva eventually let go and reach the outside world in the host's poo. The congo-floor maggot is a strange one, as it is the only fly larva that is known to suck blood from humans. The maggots lie hidden in the soil beneath sleeping mats and emerge at night to suck blood from the dozing person.

The larvae of some parasitic flies, such as screwworms, can't penetrate the host's skin, instead they gain access through a wound, even tiny ones such as insect bites. The larvae, often loads of them, feed on the host's flesh and can be very destructive.

For me, the standout parasitic insects are the *Strepsiptera*. These creatures are truly bizarre in every sense of the word and they parasitize a swathe of insects, including bees, wasps, leafhoppers, grasshoppers, crickets, cockroaches, and silverfish. One last thought in terms of parasitism is that insects have plenty of their own parasites. For example, the horsehair worms are a small group of animals where every single representative is parasitic. In addition there are parasitic nematodes, mites galore, trematodes, and some pretty disturbing fungi. Many of these are body snatchers, driving the host to strange ends so that they may complete their own life cycle.

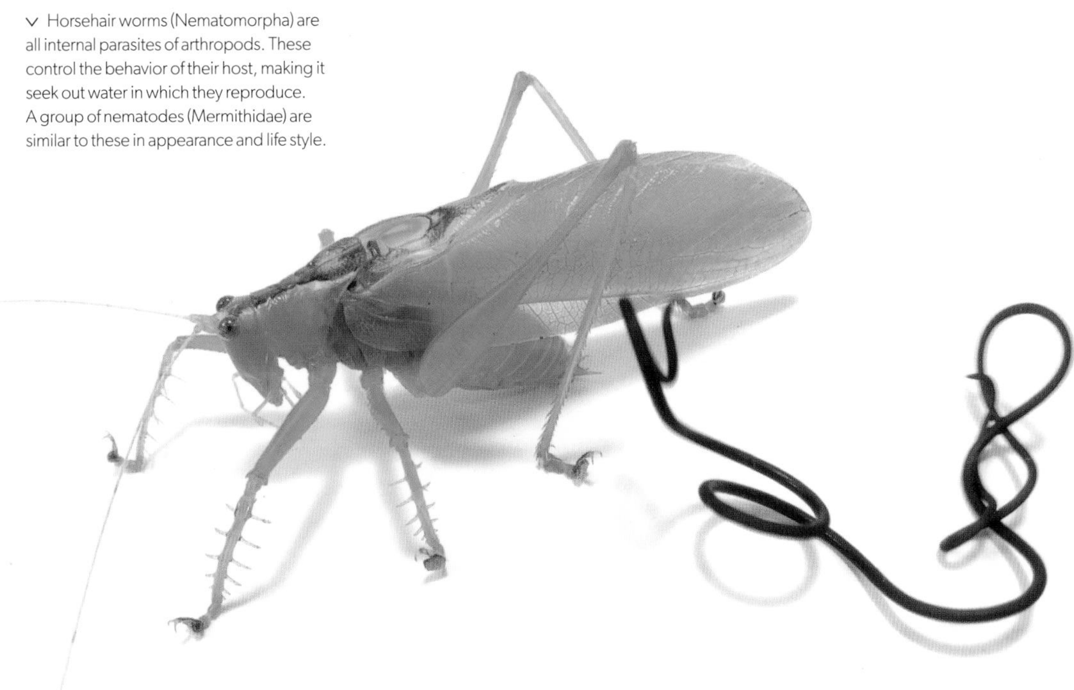

∨ Horsehair worms (Nematomorpha) are all internal parasites of arthropods. These control the behavior of their host, making it seek out water in which they reproduce. A group of nematodes (Mermithidae) are similar to these in appearance and life style.

STREPSIPTERANS

Adult male and female strepsipterans are very different in appearance, but both start out as a small, hyperactive larva, scooting freely around in the body cavity of their mother. When the time comes to leave, it exits the body cavity through a genital pore and moves down a narrow brood canal to the outside world. The female produces so many young that the vegetation where they are released becomes a bustle of squirming and jumping larvae, all eager to find a host. In the species that parasitize bees and wasps, the jostling larvae make for flowers, which may be visited by hosts foraging for nectar and pollen. When an insect, the right size, shape, and color for a bee or wasp comes within range of the larva, it uses long, stiff, paired bristles at its hind end to launch itself into the air with the hope of hitting the buzzing insect, where it clings on for dear life with its clawed feet. The bee or wasp, unperturbed by the presence of its new passenger heads back to its nest to feed its own larvae with the food it collected on its foraging trip. Once in the nest of the host, the strepsipteran larva disembarks from the ride it hitched and makes for one of the plump host in its little cell. The tiny parasite creeps along the body of the host until it reaches a spot it can burrow into. Using enzymes, the parasite larva dissolves the cuticle of the host and sinks into its body, wriggling furiously as it goes. This frantic squirming separates the various layers of the host's skin, forming a pocket that the parasite slips into. Its place inside the host now secured, the larva goes through its first metamorphosis, turning into a grub-like larva.

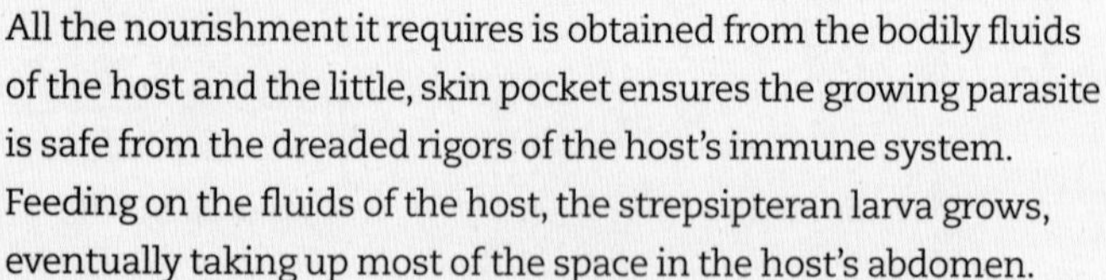

All the nourishment it requires is obtained from the bodily fluids of the host and the little, skin pocket ensures the growing parasite is safe from the dreaded rigors of the host's immune system. Feeding on the fluids of the host, the strepsipteran larva grows, eventually taking up most of the space in the host's abdomen.

The effect of this parasitism on the fully grown host is significant. The sexual organs of the adult host do not have room to mature due to the space taken up by the parasite. The host develops into adulthood, but it can be very damaged, sterilized, and sexless; and from between the tough plates of its abdomen protrudes the head end of a strepsipteran pupa, sometimes several.

˅ Male strepsipterans are very short-lived—as little as two hours in some species—and have elaborate antennae to detect the pheromones given off by adult females (*Stylops ovinae*).

Soon after the ravaged host begins its normal adult activities, the cap of the parasite pupa opens and out pops an adult male strepsipteran. His mouthparts are small and useless, and the energy reserves he built up as a larva will not sustain him for long, so he must seek out a mate as quickly as possible. His mate is nothing like him. She found her own host in the same way as the male, but she still looks like a grub and is still to be found in the host insect with her head end projecting from its abdomen. The male is attracted to the scent of his mate and copulates with her head, just behind which is the entrance to the brood canal. His sperm fertilize the female's eggs and before long a new generation of mobile larvae will be ready to begin the complex cycle all over again.

DIVERSITY AND HOSTS

- 600 known species
- *Strepsiptera* are known to parasitize bees, wasps, flies, leafhoppers, cockroaches, and silverfish
- Males are very short-lived and have little time to find a female to mate with

> A female strepsipteran (*Stylops* sp.) poking out from the abdomen of a solitary bee (*Andrena scotica*). The female parasite never leaves the host.

∨ Left: A female strepsipteran (*Stylops* sp.) within the dissected abdomen of its solitary bee host (*Andrena scotica*). The parasite takes up a lot of space in the host's abdomen.

Right: The adult females of a small number of species are free living and incredibly difficult to find (*Eoxenos laboulbenei*). As larvae they parasitize silverfish. The parasite takes up a lot of space in the host's abdomen.

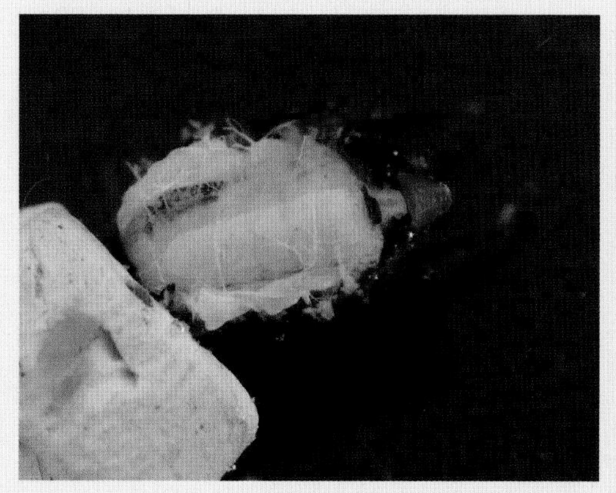

CLUMSY SUCKERS

Moths—all fluffy and cute! Aren't they? Not even moths are above parasitism. *Lobocraspis griseifusa* laps at the tears of herbivorous mammals, brushing its proboscis against the eye to induce more tears. Some moths have even become vampires. These bloodsucking moths are especially interesting because they show evolution in action, and the leading edge of an evolutionary trajectory that started with nectar feeding, went through fruit piercing, and is now at mammal piercing. Moths typically feed on nectar, which they suck through their proboscis. *Calyptra* species and their close relatives feed on all sorts of things, including fruits and the tears of large animals. In order to pierce the tough rind of fruit, these moths have evolved a sturdier and sharper proboscis. Males of ten of the seventeen described *Calyptra* species have been observed piercing mammalian skin and feeding on blood under natural and experimental conditions. Unusually among blood feeders, these moths aren't sucking blood for the protein it contains. Rather they are after the salt. Normally, male moths and butterflies "puddle," or visit sources of salt more frequently than females, and there's even some evidence that males transfer salts to the females during mating.

The ability to penetrate the tough rind of fruit is a pre-adaptation that allowed them to explore other sources of nutrients in addition to fruit, i.e. mammalian blood. These are clumsy suckers though—the switch to bloodsucking appears to be fairly recent and they're not all that good at it. The size of these moths, the way they fly, and their light color puts the wind up their hosts. They even give elephants the fear. Being so big, these vampire moths can't get amongst dense fur and the punctures inflicted by their large proboscis would be way too painful for small- and medium-sized animals. Instead, they're limited to sucking blood from large, sparsely furred mammals, such as tapirs and rhinoceroses.

∧ *Lobocraspis griseifusa* sweeps its proboscis across the eye of the host, irritating the eyeball, encouraging it to produce tears. It can even insert its proboscis between the eyelids, allowing it to feed when the host is sleeping. Moths of the genus *Poncetia* have a short proboscis, so they must cling to the eyeball itself to drink. They must be careful though. If the weeping host blinks, the moth is often crushed to death.

> A few species of moth drink the blood of vertebrates. Originally fruit feeders, using their sharp proboscis to pierce the tough rind of fruits, they appear to be relatively new to blood feeding (*Calyptra thalictri*).

6

QUID PRO QUO

Beyond the parasitoids and parasites, insects have myriad, more nuanced relationships with a huge range of other organisms. Pollination is probably the most significant of all of these. Insects move plant pollen around in exchange for nectar and some of the pollen. Many of these relationships involve microorganisms, such as the viruses harnessed as biological weapons by parasitoid wasps, and the myriad microbes in the guts of insects, which help them to make the most of their food.

INSECTS AND FUNGI

A variety of insects and fungi are completely dependent on one another, to such an extent that the insects often have little pouches that have evolved to carry the fungi to a new home.

Some beetles are farmers—a very unusual trait among animals. They excavate and maintain tunnels and galleries in wood for the purposes of cultivating fungi, which is what they eat. In the bark beetles at least, this close relationship with fungi has driven the emergence of a complex, eusocial lifestyle.

After mating, the female bark beetle (*Austroplatypus incompertus*) tunnels into the heartwood of a eucalyptus tree, a laborious process that can take seven months. Once in the heartwood, she lays her eggs, and her larvae can take from two to four years to reach adulthood. Instead of striking out on their own, the daughters remain in the nest and effectively become workers, forgoing reproduction, and helping to extend and maintain the gallery and its crop of fungi. These workers even lose sections of their legs, so they can't survive outside the nest. The nests of this eusocial beetle are far from enormous—on average they contain a single queen, five workers, and thirty-six larvae. In another species of bark beetle (*Xyleborinus saxesenii*), the larvae are the workers, cleaning and expanding the nest, while the adult females protect the nest and tend to the fungus crop.

< The "fur" of bees and lots of other pollinating insects enables them to carry more of the pollen.

> Many insects that feed on wood depend on an army of microbes in their gut to produce the necessary enzymes that digest the wood.

∧ The galleries produced by bark beetles enable them to cultivate fungi in a fully controlled environment.

< Insects and their fungal partners are crucial in forest ecosystems—injuring and killing trees to make space in the canopy and returning dead trees back to the soil.

In ship-timber beetles (*Lymexylidae*) it is exclusively the larvae that cultivate fungi. The female deposits eggs in crevices and cracks in the wood, using her ovipositor. As each egg is laid, she coats it with fungal spores from a little pouch near her ovipositor. The larva hatches and carries some of the fungal spores with it as it chews its way into the wood. The tunnel that the larva excavates is lined with a white layer of the fungus and this is what it eats. The larva takes excellent care of its fungus garden doing everything it can to keep the conditions just right for the fungi, so that it will have enough food to complete its development. The fungus requires oxygen to thrive, so the larva must rid the tunnel of any debris to maintain a good flow of air. The larva shuffles along the confines of its tunnel pushing any wood dust and waste to the outside, where it falls to the base of the tree.

The success of many other insects, such as leafcutter ants (see pages 130–133), also owes a great deal to fungi. The ants build a subterranean metropolis that is effectively a fungus farm. In return for feeding the fungus and giving it a nice, stable, competitor-free environment to live in, the ants eat specialized bits of the fungus, which are called gongylidia.

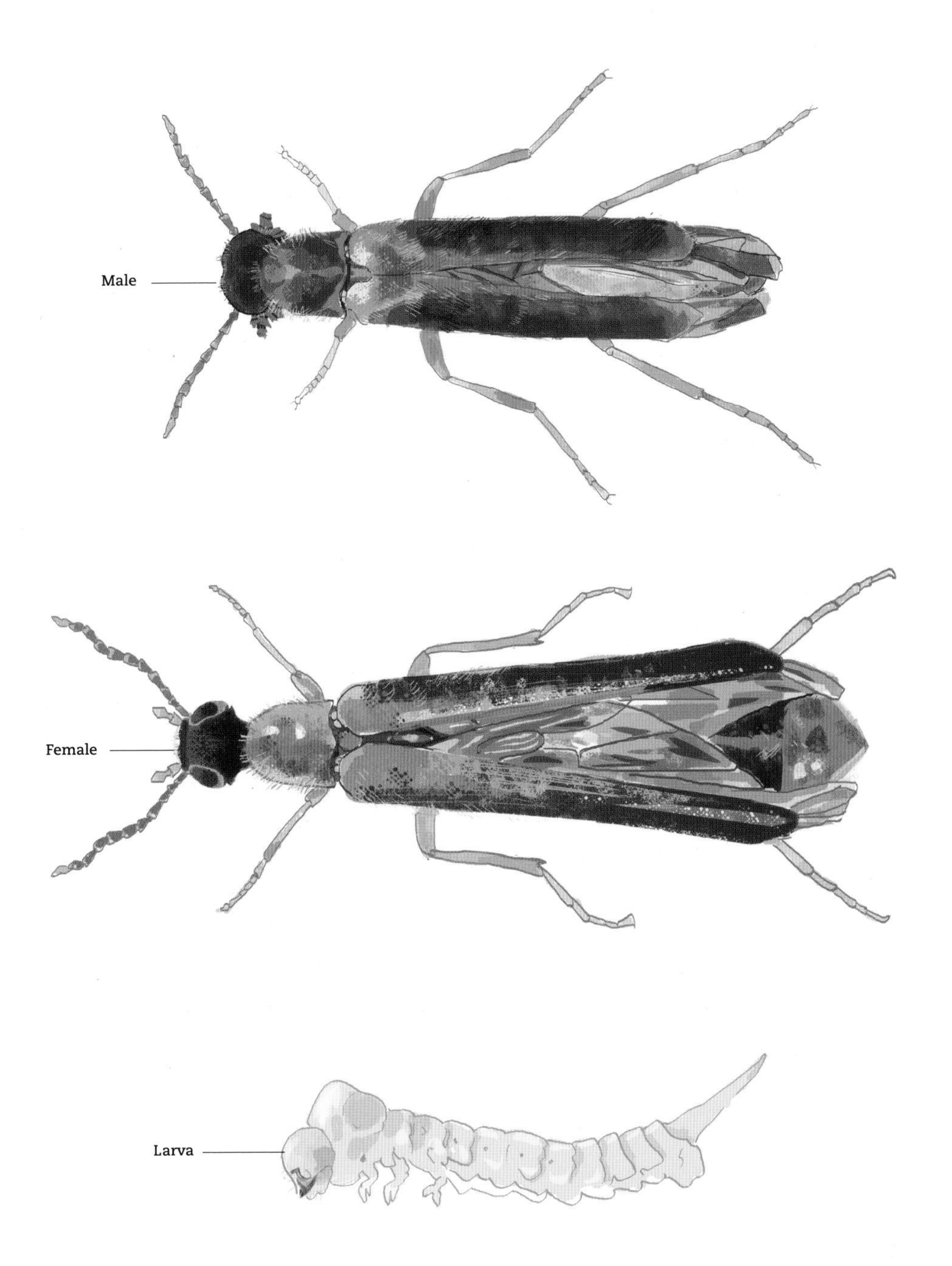

^ Ship-timber beetles have a mutualistic relationship with a type of yeast. Their larvae excavate tunnels in dead wood where they cultivate the yeast to eat.

INSECTS AND MITES

There are lots of intriguing interactions between insects and mites, but on the whole very little is known about what is exactly going on. In the few relationships that have been studied in any detail, some remarkable insights have been gained. Some male solitary wasps appear to use mites like a chastity belt, and some beetles wear an insulating jacket of mites.

The males of some solitary wasps are infested with mites, which cling to their abdomens and scuttle aboard the female during mating, making straight for her genital opening to block it up. This prevents her from mating again, thus securing the male's paternity. This is extraordinary, yet it is only the tip of the iceberg, because the mites have to get onto the next generation of wasps, so they stay on the female while she builds a nest. This nest is usually inside a hollow plant stem, and it eventually consists of a number of cells with mud partitions, each of which is stocked with small, paralyzed caterpillars—food for the wasp's offspring—and a single wasp egg.

During the construction of the nest, the mites (immature at this point) leave the female and initially feed on the hemolymph of the poor, paralyzed caterpillars. The wasp larva, once hatched, makes short work of the provisions, grows quickly, and pupates.

< The nests of insects, solitary and social species alike, are home to a huge range of mite species. On the whole, little is known about the exact interactions between the mites and their hosts. In some potter wasps, there is evidence of some fascinating interactions (*Ancistrocerus nigricornis* nest).

Here, there is yet another astonishing twist in the tale because the fate of the mites in each cell of the nest depends on whether the wasp larva is male or female. Once a female wasp larva has finished eating the paralyzed caterpillars, it seeks out and consumes every single mite in its cell before it pupates. In contrast, the male wasp larva leaves the mites in its cell alone and pupates. The now adult mites shift their attentions to the male wasp pupa and suck its hemolymph. These sexually-transmitted mites now mate and have their own offspring that also feed on the male wasp pupa.

When the adult wasps finally emerge, the new batch of immature mites hidden in the cells of the male wasps climb aboard their new hosts to continue this extraordinary cycle. As the female wasps leave the nest, they may occasionally pick up a couple of mites from the empty male cells as they head to the exit, but on the whole they are mite-free.

∧ This unidentified *Parancistrocerus* potter wasp from Taiwan has a close association with mites, the exact nature of which is unknown. The abdomen shown in the image below is of this species.

∨ In some wasps, mites are sexually transmitted from male to female during mating. The abdomen of these wasps even has a special recess—the acarinarium—where the mites pack together. These mites are on the abdomen of an unidentified female potter wasp from Taiwan.

^ Many insects are often infested with mites. In the burying beetles at least it seems the mites may work like an insulating jacket, helping them to get active and fly before rivals with no or fewer mites.

Since this extraordinary symbiosis was first investigated more than thirty years ago, no one has really looked into it, and there are many questions still to explore. At face value, it would appear that the males, in transmitting the mites, are reducing the fitness of their offspring, but we have to remember the mind-bending haplodiploid sex determination of wasps. In these animals, fertilized eggs develop into females and unfertilized eggs end up as males. Therefore, only the female offspring have any of their father's genes. In contrast, the male offspring derive all of their genes from their mother.

The mites are getting a lot out of this relationship. They get dispersal, food (the hemolymph of the caterpillars and male wasp pupa), and a safe place to develop (the wasp's nest). What do the wasps get? The mites act as a post-coital chastity belt for the male wasps—guaranteeing paternity—but that's all that is known. The male wasps must pay a price in terms of fitness because the mites feed on their hemolymph. Perhaps a male wasp with a heavy burden of mites is more ravishing to female wasps? Perhaps the mites feed on parasitoid eggs and larvae inside the nest? Perhaps the female wasp larvae gain something from consuming all of the mites in their cell? There's still so much to understand and if you're a biologist just starting out this would be a fascinating world to explore.

INSECTS AND MICROBES

Microscopic, single-celled organisms are everywhere and it's no surprise that insects have forged some remarkable relationships with them. There are viruses that are used as biological weapons, bacteria that are used as antibiotics, and a hoard of microbes that help insects digest their food.

Biological Warfare

There are parasitoid wasps that have developed a fascinating symbiosis with viruses, this relationship goes back around 100 million years. The DNA of the virus is actually in the wasp, dispersed through its genome, and it only replicates in a particular part of the female wasp's ovary, known as the calyx. The viruses are injected with the wasp's eggs, like microscopic bodyguards, interfering with and disabling the host's immune responses that would encapsulate and kill the eggs. Free from the attentions of the host's last line of defense,

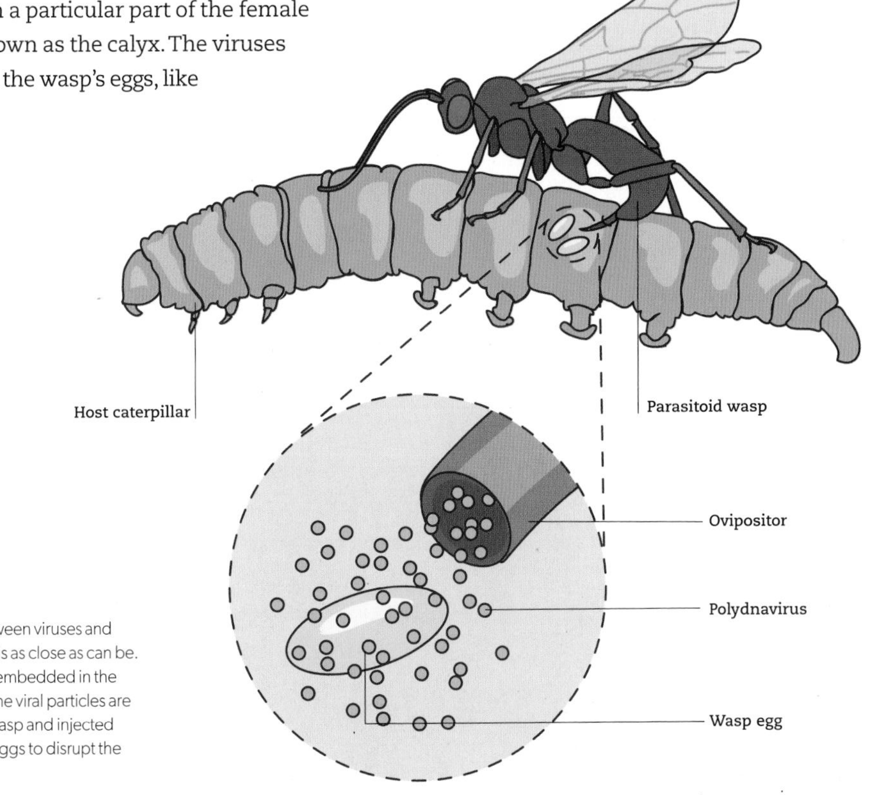

> The relationship between viruses and some parasitoid wasps is as close as can be. The DNA of the virus is embedded in the genome of the wasp. The viral particles are assembled inside the wasp and injected along with the wasp's eggs to disrupt the host's immune system.

the egg hatches and the wasp larva feeds on the soft, juicy insides of the host, which is often a caterpillar.

As well as defeating the host's immune system, the virus also tweaks the metabolism and development of the host, making conditions just right for the growth and survival of the wasp larva. In the species that have been studied in detail, the host continues as normal, but when it has shed its skin for the fourth or fifth time, it pupates prematurely and dies, by which point the wasp larvae are mature and ready to pupate.

> Overwintering insects are at risk from fungal and bacteria pathogens, which has driven the evolution of some fascinating countermeasures (*Philanthus triangulum* emerging from pupal cocoon).

∨ From special cavities on her antennae, a female beewolf secretes a white paste of bacterial cells (*Streptomyces philanthi*). These bacteria inhibit the growth of pathogens that might otherwise destroy the wasp's developing young.

Bacteria and Antibiotics

The beewolf (see page 55) has also forged a remarkable alliance with microbes—namely bacteria. This solitary wasp provisions a subterranean nest with paralyzed honey bees. However, before she seals the brood chamber, the beewolf leaves a parting gift. From little cavities on her antennae, she daubs some blobs of white matter onto the interior wall of the chamber. This secretion is actually a mass of symbiotic bacteria cells (*Streptomyces philanthi*) that lives with the beewolf and nowhere else.

The symbiotic bacteria have two clever functions. Firstly, the little white mass of bacteria functions like an exit sign in the pitch black of the brood chamber, guiding the position in which the larva pupates, so that its head faces the main tunnel, ready for when it emerges as an adult the following summer. Correctly positioned, the larva spins its silken cocoon, incorporating some of the symbiotic bacteria as it does so. This is the second function of the bacteria, as they turn the cocoon into an antimicrobial fortress by producing compounds that kill harmful bacteria and fungi that might otherwise infest the inactive larvae during the wet, colder months.

Partners in Digestion

Microorganisms are actually a fixture of the digestive tract of all animals. These gut organisms—collectively known as microbiota—aren't just bacteria either, they are fungi, protists, archaea, and viruses. This is an exciting area of biology because it seems that these gut organisms are important to the host animal in many ways. In insects, as with other animals, the gut microbiota have been shown to contribute to the host's digestion, detoxification, development, pathogen resistance, and overall physiology. Behind the success of every insect that thrives on a restricted diet, there will be an army of gut organisms producing nutrients from the food with tricks embedded in their hyperdiverse genomes.

Some of the best-studied insect microbiota are those found in the guts of termites. Spare a thought for these much maligned insects as they're remarkable creatures. Effectively, social cockroaches, the ecological importance of termites is matched by few other insects and their success hinges on their ability to make short work of that most-refined delicacy—wood. Wood is unlikely to be on the menu at any restaurants in the near future, but it is loaded with energy if you can break apart the constituent molecules. The microbes in termite guts do this with aplomb, putting these animals right up there with fungi in terms of their ability to degrade wood. This doesn't sound like much, but it's not something to be sniffed at. On land, this is a crucial ecological process as it recycles the nutrients and energy locked up in woody tissue.

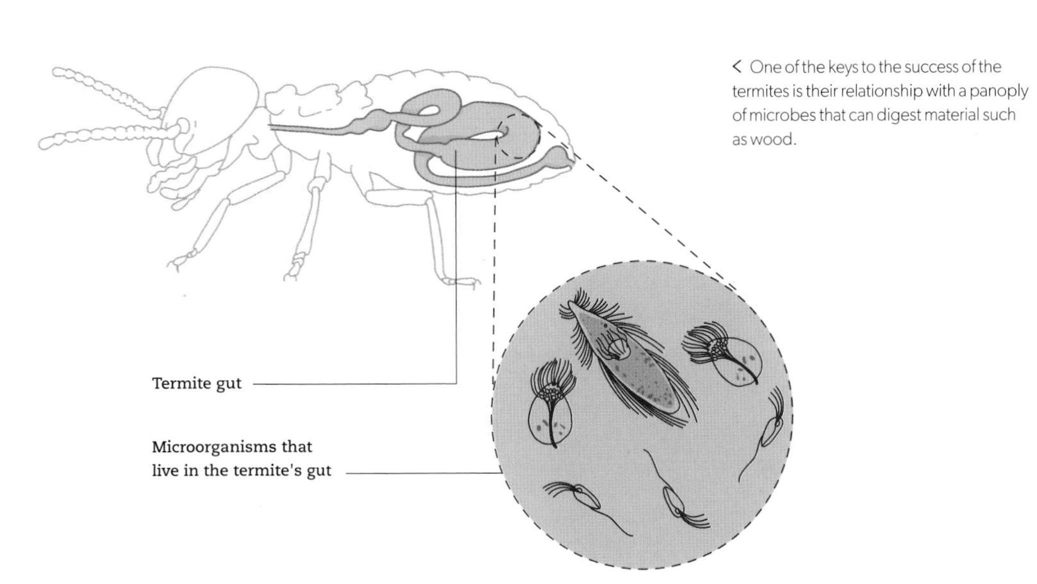

< One of the keys to the success of the termites is their relationship with a panoply of microbes that can digest material such as wood.

INSECTS AND PLANTS

To find some of the most astonishing interactions between insects and other organisms we must look to those that have sprung up among insects, plants, and sometimes a third party.

The Assassin Bug and *Roridula*

The assassin bug (*Pameridea roridula*) has formed a remarkable alliance with a plant, *Roridula*, which lives on impoverished soils. Other plants unable to find sufficient nutrients in the soil have become carnivorous and are able to catch and digest insect prey. Just think of the well-known Venus fly trap or the pitcher plant. *Roridula* plants can catch insects using sticky hairs, but they do not produce the necessary enzymes to digest them, so they have formed a strange alliance with an assassin bug. The bug feeds on the insects that are trapped by the plants' sticky hairs and when they defecate they do so on the plant, allowing it to absorb the nutrients that are otherwise inaccessible to it.

∨ The small assassin bug *Pameridea roridula* is something of a middle man in the life of the "carnivorous" *Roridula* plant. The bug feeds on insects trapped by the sticky hairs of the plant. When it defecates the nutrients are absorbed by the plant.

Ants and Plants

Ants have some of the most fascinating and complex relationships with plants. There are at least 100 species of ant that live in a close partnership with a plant. These exemplify the degree to which insects and flowering plants have become inextricably linked over millions of years. In return for the services provided by the ant, the plants provide them with accommodation. The diversity of the relationships encompasses all the conceivable parts of a plant. Some plants have modified swellings on their branches and twigs, while others have cavities within their stems and trunks. Yet more have modified roots that house their insect guests.

In South America, a few species of tree (*Cordia nodosa*, *Tapirira guianensis*, *Duroia hirsuta*) dominate small areas of the rain forest, forming patches that the local people call "Devil's gardens." Within the hollow stems and the leaves of these trees there are special cavities in which lemon ants (*Myrmelachista schumanni*) makes their nests. Any saplings of other plants sprouting near the host trees are attacked by the ants and stung with formic acid, which kills them, thus freeing the host trees of their competitors. Any animal attempting to nibble the leaves of the trees is treated in the same way.

> Greater celandine (*Chelidonium majus*) seeds are dispersed by ants. The little waxy things attached to the seeds are elaiosomes, which ants like to eat.

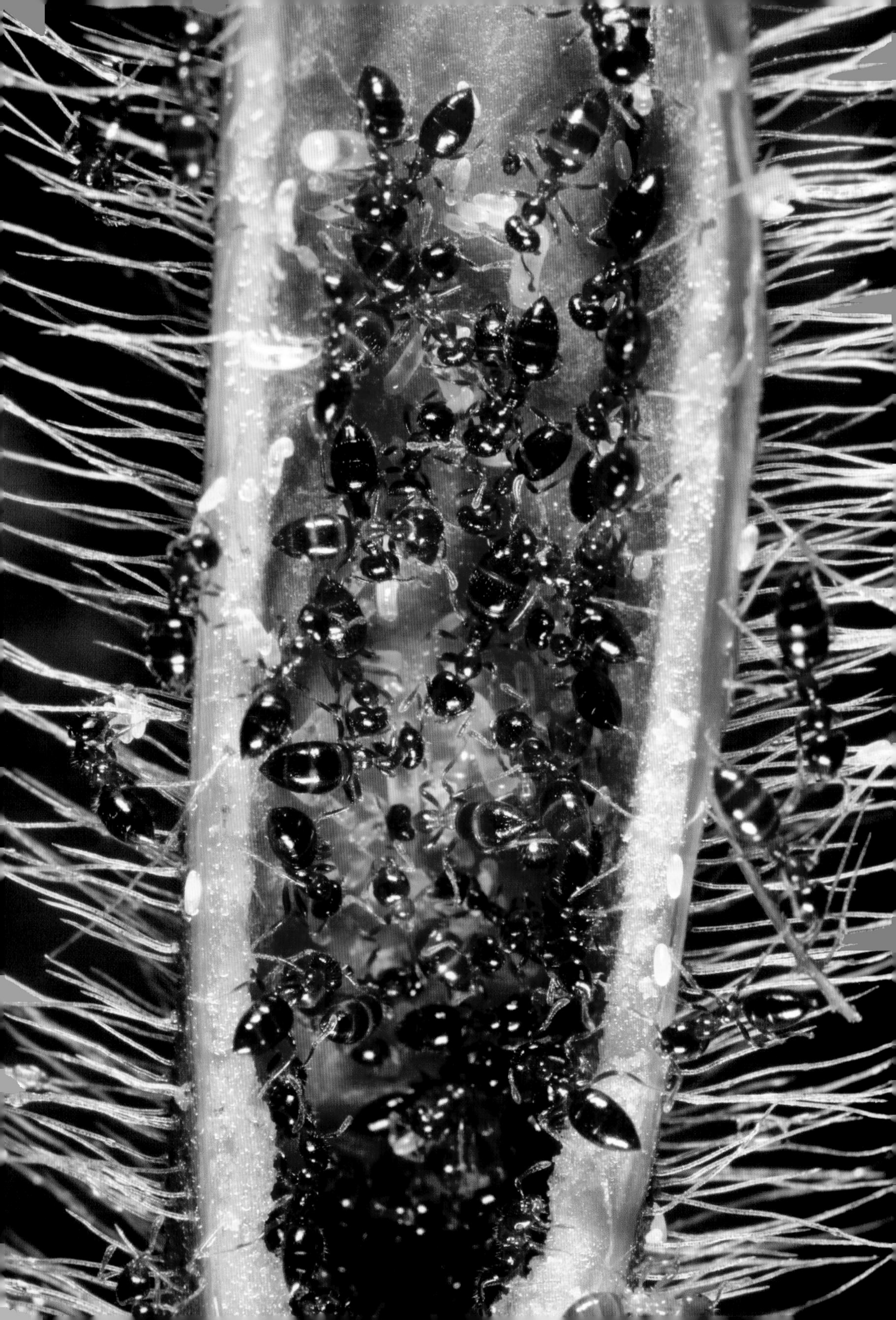

∧ Swellings on the plant, *Cordia nodosa*, are inhabited by ants. The ants will damage and destroy rival plant species.

< Ants (*Myrmelachista schumanni*) nesting in a branch of the tree *Duroia hirsuta*.

> The bulbous base of *Myrmecodia* plants are living quarters for ants. The plant has no roots, instead it absorbs nutrients from the ants' waste.

Myrmecodia are strange epiphytic plants that also have a remarkable partnership with ants. These plants cling to the trunks and branches of trees with no roots in contact with the soil, and it is their strange bulbous tubers that house the ants. Open up one of these tubers and you'll see an elaborate system of tunnels and chambers. Some of these chambers are the nest's rubbish tips and the waste therein is used by the plant as a fertilizer, allowing it to flourish even though its roots will never come into contact with the soil.

Bull Horn Acacias

Bull horn acacias also have an intriguing relationship with ants, which begins when a young, newly mated queen ant lands on the acacia looking for a place to start a nest. The queen, convinced by the odor of the tree that she is in the right place, starts to nibble a hole in the tip of one of the bulbous, hollow thorns, eventually breaking through to the cavity within. Inside she lays her first batch of fifteen to twenty eggs and over time the embryonic colony grows and expands into more of the hollow thorns.

When the colony has exceeded around 400 individuals, the repayment to the acacia for lodgings can begin and the ants assume their plant-guarding role. The ants become aggressive and have a go at any creature munching the acacia's leaves, regardless of whether it is a cricket or a goat. It doesn't take much to set them off. Even the whiff of an unfamiliar odor sees the ants swarming from their thorns and toward a potential threat. Herbivorous insects are killed or chased away and browsing mammals are stung in and around their mouth, which quickly persuades them to look elsewhere for less well-defended fodder.

Apart from these active defending duties, ants also have gardening to tend to, so they leave the tree and scout around its base looking for any seedlings that would eventually compete with their acacia for light, nutrients, and water. If

∨ Beltian bodies are little packets of protein and fat produced by some *Acacia* species as a "reward" for their ant cohorts.

∧ Thorns, leaves, and Beltian bodies on a bull horn acacia tree (*Vachellia cornigera*). The relationship between these trees and their associated ants are among the best studied plant–insect symbioses.

they do find any, they destroy them, and even go so far as to prune the leaves of nearby trees, so their host does not suffer from too much shade. It also seems as though bacteria living on the bodies of these acacia ants might also protect the tree from harmful bacteria.

Not only does the tree supply the ants with nesting sites, but special glands at the base of the tree's leaves produce a nectar rich in sugars and amino acids that the ants lap up. The tips of the leaves also sprout Beltian bodies. These are small, nutritious packets of oils and proteins, which the ants snip off and carry away to feed to their grubs. The grubs even have a little pouch at their head end that the Beltian body can be tucked into while they feast on it.

Pitcher Plants

In the steamy forests of Borneo, a species of pitcher plant has struck up a multifaceted relationship with an ant species. The ants live in the base of a hollow tendril and from there they make regular forays into the plant's pitcher to drag out larger insects floundering in the liquid, even swimming to get at the choicest morsels. The ants haul the half-drowned victims up under the rim of the pitcher and consume them. Pieces of the prey and the ants' feces fall back into the pitcher and are digested. Who's getting what here? The ants are getting easy meals, sugary rewards from the plant's nectaries, and a place to live, but it seems the pitcher plant is just getting robbed. It turns out that larger prey items in the pitcher can rot before they're digested, thus killing the pitcher. In essence, the ants are acting a bit like the plant's digestive system, breaking the prey into smaller bits that the plant can cope with. The ants also keep the rim of the pitcher clean, maintaining its slipperiness, so that other insects fall into the trap more easily.

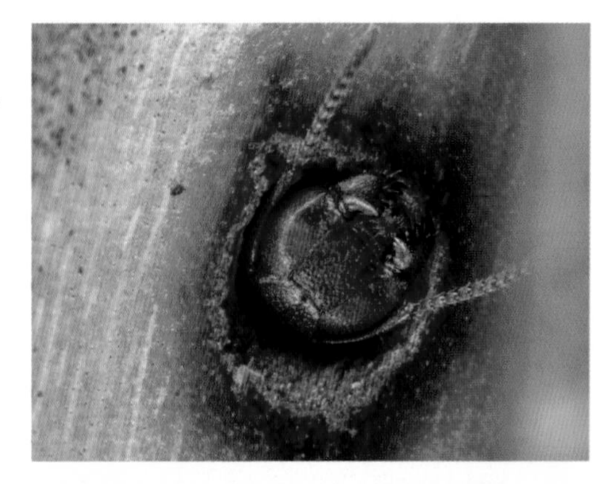

∨ > The diving ant (*Camponotus schmitzi*) has a complex relationship with the pitcher plant *Nepenthes bicalcarata*. The ants effectively help the plant digest larger prey items that fall into its pitcher.

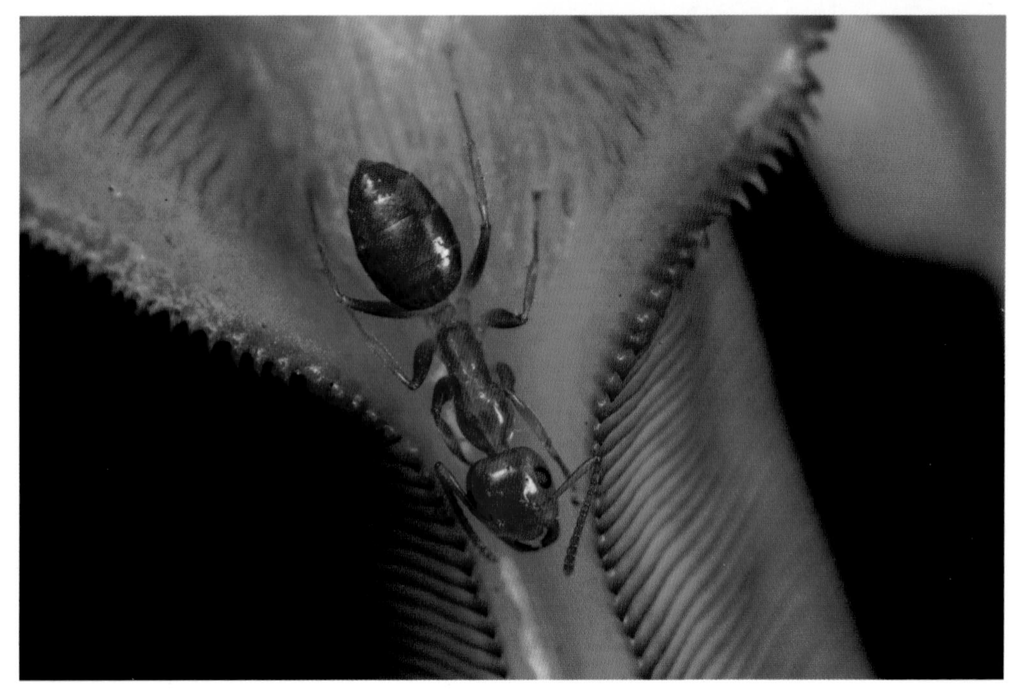

^ Many insects, especially ants have a real liking for honeydew, the sweet liquid produced by sap-sucking insects, such as this scale insect (*Paralecanium* sp.). They like the stuff so much they even protect and shepherd the sap-suckers.

Ants, Plants, and Sap-suckers

Aphids, whiteflies, scale insects, mealy bugs, treehoppers, cicadas, and leafhoppers are all sap-sucking insects. Sap is mostly water, so all of these animals must get rid of the excess fluids as they're feeding. Droplets of sugary fluid—honeydew—emerge from the sap-sucker's back end. Many animals have a special taste for this sugary treat, none more so than the ants. Honeydew makes up at least 90 percent of the diet of some ant species and they like it so much they treat these sap-suckers like cows—herding them, protecting them, and milking them. An ant delicately strokes an aphid with its antennae to elicit a droplet of honeydew that it greedily consumes. These miniature herders are very protective of the sap-suckers and will drive off the predators and parasitoids that would otherwise make short work of them. They'll even move the herd to better feeding grounds to increase the production of honeydew. This relationship is one of the reasons why some sap-sucking bugs are so successful.

ANTS, PLANTS, AND FUNGI

One of the strangest ant–plant mutualisms also involves a fungus. Beneath the branches of its host tree (*Hirtella physophora*), the trap ant (*Allomerus decemarticulatus*) constructs small galleries, which look like pockmarked areas of dense gray webbing. The ants construct these galleries using hairs stripped from the surface of the host tree, their saliva, and a specially cultivated fungus that acts as a resin, making a sort of fibrous matting—essentially the insect equivalent of fiberglass. These galleries have been known about for a long time, but it was thought they were simply a refuge for the ants while they were away from the nest foraging for food. As it turns out, the real purpose of these galleries is much more macabre.

The worker ants amass in these galleries with their little heads just inside the holes, their powerful serrated jaws agape. Here, they wait in ambush and before long a large, plump cricket ambles in to view exploring the branch tentatively with its long limbs and sensitive antennae. The hairs that cover the tree's outer surface are a deterrent to walking insects. The smooth surface of the gallery on the other hand will seem like a fine place to rest or have a snack. Sensing nothing out of the ordinary about the inconspicuous gallery, the cricket walks straight onto it. Two or three of its clawed feet may probe the holes for purchase and it is then the waiting ants strike. They grab the ends of the cricket's legs and heave, pinning the cricket to the surface of the gallery. Other workers rush out and begin dragging the remaining limbs and antennae into the holes until the prey is well and truly snared. Other workers then swarm all over the prey and sting it.

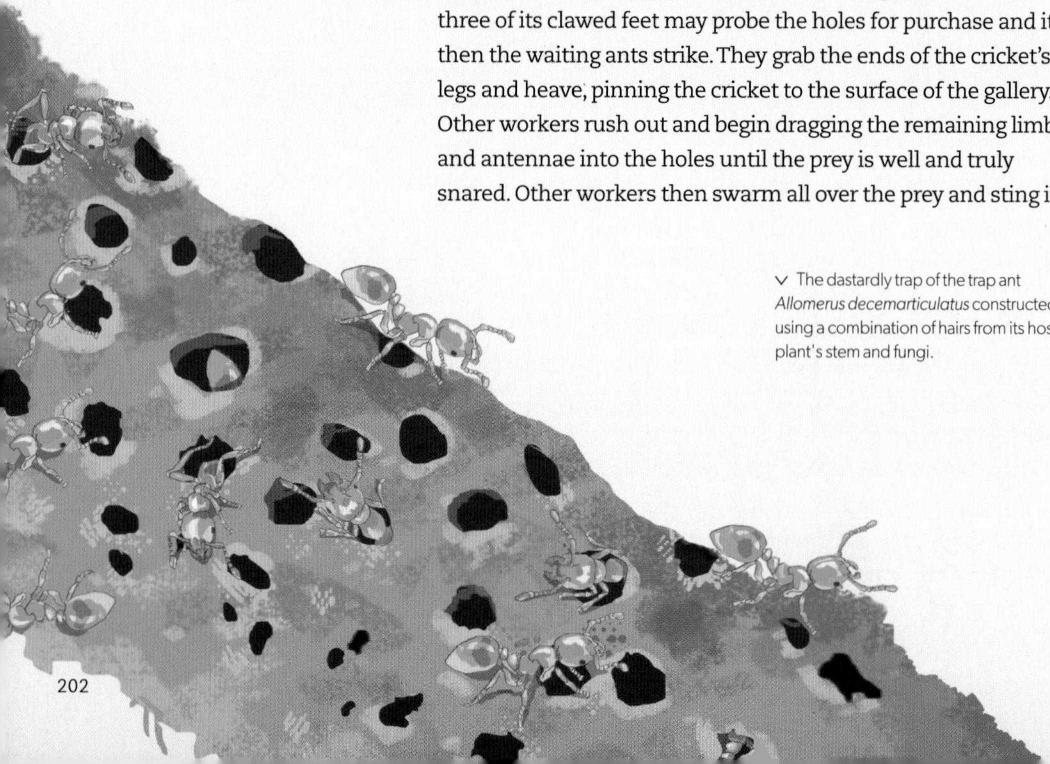

∨ The dastardly trap of the trap ant *Allomerus decemarticulatus* constructed using a combination of hairs from its host plant's stem and fungi.

With the prey immobilized and at death's door, the ants can begin their gory work. Using their tenacious jaws, they begin butchering the carcass of the cricket, chopping out chunks of flesh to carry back to the nest pouches, where the developing ant's young can be found. This whole strategy is the product of a symbiotic relationship between the ant, its host tree, and the fungus. The ant has a safe place to nest in the leaf pouches of the tree. A favor repaid by ridding the tree of herbivorous insects that can't wait to get their mandibles into the succulent leaves of the *Hirtella* tree. The fungus is used as an adhesive in the construction of the traps, but in being used this way it is cultivated by the ants and spreads to areas that it might not reach otherwise.

∨ Larger insects blundering onto the trap, like this grasshopper, are pinioned by the ants and then killed and butchered. The remains of the grasshopper have attracted a wasp.

^ Stingless bee (*Oxytrigona* sp.) attending to treehopper nymphs (*Aethalion reticulatum*) and taking a droplet of honeydew.

More Than Meets the Eye

The relationships between insects and other living things are often more complex than they appear at first sight, reflecting the "power struggle" that exists between the species in these interactions. For example, the bull horn acacia's ants will repel most herbivorous insects, but they turn a blind compound eye to the sap-sucking antics of scale insects, which suck the sap of the host acacia, thus weakening it and providing entry for disease. The ants tolerate and even protect the scale insects because they produce sweet honeydew that they relish.

Cordia nodosa trees in the Amazonian rainforest have a kind of partnership with *Allomerus octoarticulatus* ants, which make their nests in modified leaves. These ants will actively destroy the tree's flower buds to increase the amount of living space available to them. The flowers die and leaves develop instead, provisioning the ants with more dwellings. A similar thing can happen between some *Hirtella* trees and their

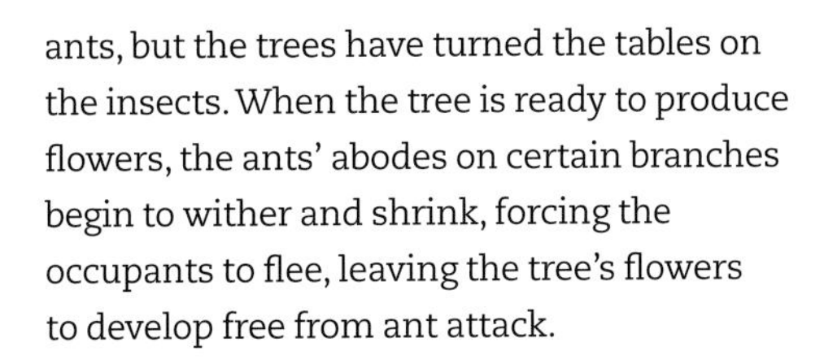

ants, but the trees have turned the tables on the insects. When the tree is ready to produce flowers, the ants' abodes on certain branches begin to wither and shrink, forcing the occupants to flee, leaving the tree's flowers to develop free from ant attack.

One group of rove beetles (*Amblyopinus* spp.) were once wrongly assumed to be blood-feeding parasites of rodents and marsupials. However, these beetles are now known to be beneficial to their hosts, actually ridding the mammals' nests of parasites, such as fleas and mites. They can be found clinging to their host mammals, often just behind the ears or at the base of the tail. If they wind up on the wrong host though they quickly end up dead.

∨ *Amblyopinus* rove beetles were long thought to be parasites on their mammalian hosts, but it seems they actually benefit the host by feeding on fleas and mites.

INSECTS IN A CHANGING WORLD

Are Insects Declining?

GROWING UP IN THE 1980s, I have vivid memories of snowstorms of moths in the headlights of our car when we drove at night, huge numbers of dung beetles being drawn to the light from our living room window, and violet ground beetles under most of the rocks in our yard. I don't see these things anymore.

Studies from around the world seem to show that insect populations are declining across the board, with some groups of insects more affected than others. Sifting through the sensationalism, these stories were prompted by a number of studies that have shown a big decline in insect and other arthropod populations. For example, a 77 percent decline in flying insect biomass was reported in sixty-three nature reserves across Germany between 1989 and 2016. Another, more recent study from Germany, reported substantial declines in grassland and forest insects. Likewise, in Puerto Rico, arthropod biomass declined ten to sixty fold between 1976 and 2012. A study in the Netherlands reported that butterflies have declined by at least 84 percent over the last 130 years.

∨ The first time I found a violet ground beetle I was transfixed. This is still a widespread species, but likely nowhere near as abundant as it once was.

The populations of some large conspicuous insects have certainly plummeted, but the situation for smaller, less well-known insects is not well understood. There are many nuances in these declines that are absent from the media reports.

Most studies of this type only deal with the large, conspicuous insects. This is because creatures like butterflies, moths, and dragonflies are easy to find, and have been studied for a long time, especially in Europe and North America. A study of butterflies in the Netherlands may tell us a lot about a few insect species in a part of northern Europe, but what does it tell us about beetles that spend most of their lives nibbling wood deep inside dead and decaying trees in Borneo? As tempting as it is, we can't extrapolate data like this to other types of insect and other parts of the world. Similarly, if an enormous decline is based on two data points separated by nearly forty years, how can we be sure that insects weren't unusually abundant in the first sample, perhaps because of a natural phenomenon, such as a periodical mass flowering event that provided a glut of food? Relying on biomass to infer a decline without identifying all the species has pitfalls, too, because rather than an overall decline we may just be seeing the decline of a small number of extremely abundant species.

Regardless of the headlines, insects are not going anywhere fast, and to even suggest that they're careering toward extinction is misconceived. We're talking about a hyperdiverse group of animals that has been around for at least 400 million years and has come through all of the catastrophic extinction events that have hammered life

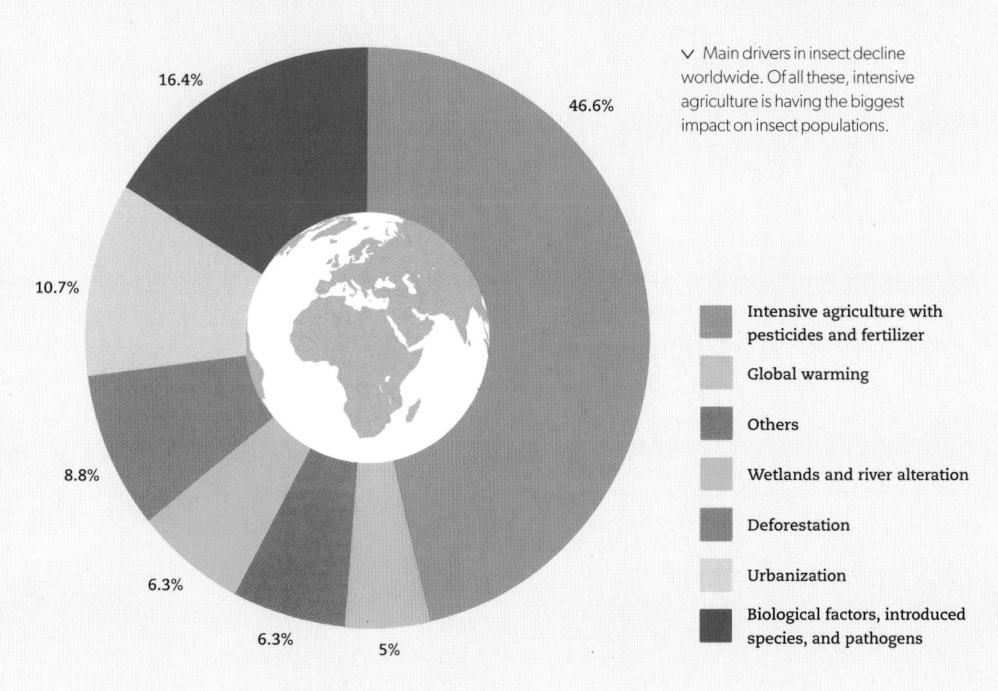

∨ Main drivers in insect decline worldwide. Of all these, intensive agriculture is having the biggest impact on insect populations.

on Earth during that time. Its diversity in form and life style, living out their lives in ways which we can barely conceive, makes them incredibly resilient, and there's nothing that humans could do that would wipe them out completely. Indeed, they'll be around long after we've disappeared.

Even though insects are not becoming extinct, there's plenty to be concerned about, especially as these animals are a barometer of the health of the environment and an early warning that all is not well. We ignore these warning signs at our peril. Insects are the linchpins of terrestrial and freshwater ecosystems. In browsing, predating, scavenging, and getting eaten by other organisms, they profoundly affect the movement of nutrients and energy through these systems. In myriad subtle ways they keep life on Earth ticking over.

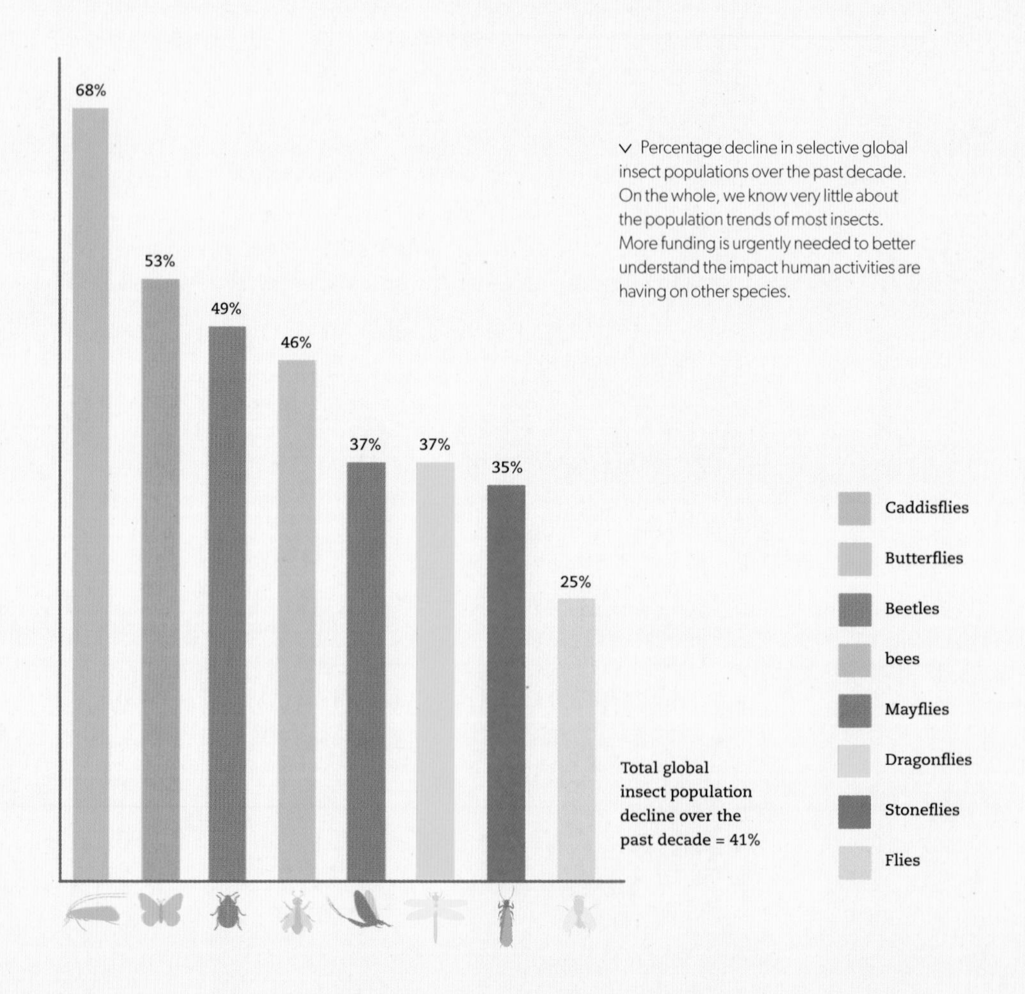

∨ Percentage decline in selective global insect populations over the past decade. On the whole, we know very little about the population trends of most insects. More funding is urgently needed to better understand the impact human activities are having on other species.

^ As climate change bites, wildfires are becoming more frequent and ferocious. These are a fixture of some habitats, but in others they can cause utter devastation.

What Is Causing Insects to Decline?

The reasons behind these insect declines can be difficult to explain, but habitat loss, the intensification of agriculture, especially the large-scale use of pesticides, climate change, light pollution, invasive species, and electromagnetic radiation are the most important factors. Human activity is simultaneously depriving insects of habitats, poisoning them, and creating a warmer world that puts beleaguered populations under even more strain.

The unrelenting destruction of diverse habitats means that we are losing species before anyone has had a chance to describe them or to understand their place in the web of life. We douse the land with a complex cocktail of chemicals, when there's only the most meager understanding of how each individual chemical affects a few organisms. Human-induced climate change is warming the planet, forcing species to move, adapt, or become extinct, and subjecting them to more extreme weather. Not only that, but seemingly small increases in temperatures can dupe overwintering insects to emerge too early and even halve the fertility of male insects.

Understanding and Reversing the Declines

The fact is though that we just don't know how insects as a whole are reacting to these pressures. Compared to vertebrates, our knowledge of insects is pitiful. Around one million species of insect have been described and there are millions more out there that await description. The vast majority of the described species are nothing more than a name, and precious little is known about how they live, not to mention their long-term population trends.

Anyone can get out there and fill in these gaps, the only things you need are time and patience. To see how curiosity and careful observation lead to stunning discoveries, just take a look at any of the fascinating insights in this book—each and every one of them rooted in curiosity and gleaned by careful observation, often over many years.

We need to embrace natural history as an important part of science. We also need to nurture the unbridled curiosity we see in children—encouraging an appreciation of life in all its glorious forms. I've yet to meet a child who isn't interested in nature, especially insects, so we should do whatever we can to show them the value and delight of observation and discovery. Insects are perfect for this as they're small and they're everywhere. Seeing lions hunting on the African savanna may be out of reach for most children, but equally compelling struggles involving much stranger, albeit smaller beasts are taking place in every backyard.

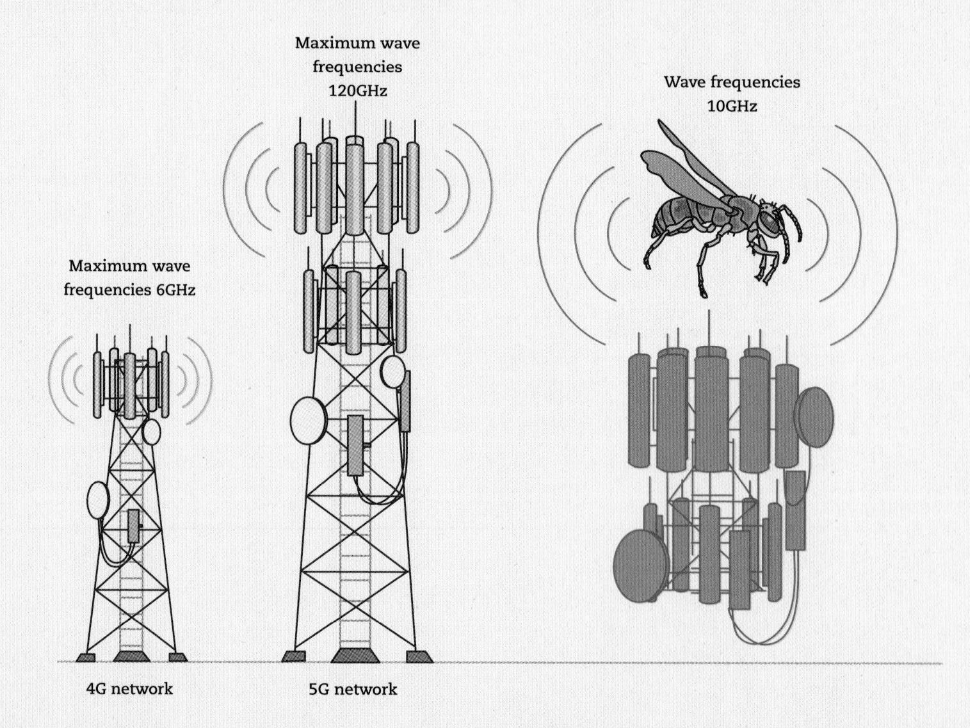

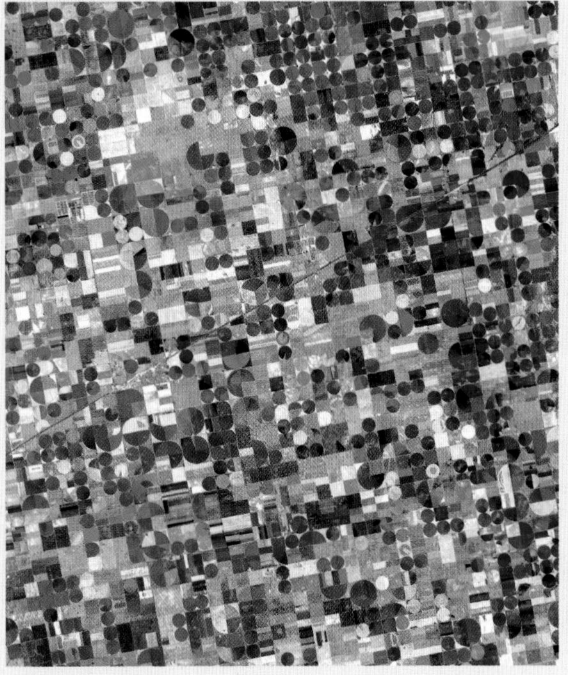

< There's mounting evidence that radiowaves and microwaves we use for telecommunications have an impact on insect populations, possibly by interfering with their navigational abilities. Wave frequencies over 10GHz appear to have an impact. The 5G network can generate wave frequencies up to 120GHz! We don't really know how this might affect insects in the future.

^ In the last forty years alone, the Earth has lost a huge amount of forest cover. Forests, particularly those in the tropics have unparalleled insect diversity.

> Conversion of land for agriculture and the associated use of pesticides is having an impact on insect populations across the globe.

FAQs: THESE ARE SOME OF THE QUESTIONS THAT ENTOMOLOGISTS ARE OFTEN ASKED

How long do insects live?

This varies enormously. Some insects, especially aphids and flies can complete their entire life cycle in a few days. The adults of mayflies and many other insects are famously short lived, but the larvae/nymphs of these often take more than a year to reach maturity. In some circumstances, beetle larvae that feed on wood can take decades to complete their development.

How do insects navigate?

Diurnal insects orientate themselves using the position of the sun as a reference point. Even on cloudy days they can still do this as they're able to see polarized light and detect the angle of the sun. At night, insects can use the position of the moon and at least one species of dung beetle (*Scarabaeus satyrus*) uses the Milky Way as a reference point to keep itself on track. Insects that have to find their way to and from a nest of some sort also use visual landmarks, for example, rocks, stones, or other conspicuous features near the nest entrance.

Do insects feel pain?

All living things can respond to damaging stimuli. Whether insects feel pain or not is a difficult question to answer. Pain is a personal subjective experience that includes negative emotions and they're unlikely to experience pain as we understand it. With that said, insects are living things and they deserve our respect.

Why do insects come to lights at night?

It is thought insects do this because they use the position of the moon as a reference point to orientate themselves. In essence, a light is an artificial moon. There's so much artificial light at night now that this is a real problem for nocturnal insects and it is thought to be an important factor in some of the population declines that have been seen.

How do insects see the world?

We don't really know. We do know the structure of insect eyes and what the different elements do, but this tells us little about how an insect brain processes the light information and creates an image. We know that insects often have more light-sensing cell types than mammals. For example, humans have three types of light-sensing cells, one for red light, one for blue light, and one for green light. Some insects have been shown to have far more, including cells that sense UV light. The common bluebottle butterfly, *Graphium sarpedon* has fifteen different types of light-sensing cells, but it's not yet known if they have better color vision than a human. It might be that some of these cell types are for detecting fast-moving objects against the sky, or particular light wavelengths reflected from potential mates or enemies. The compound eyes of insects also seem to be superb at light gathering and detecting movement. Some nocturnal insects have remarkable, seemingly impossible low-light vision. The nocturnal sweat bee *Megalopta genalis* navigates to and from its nest—a hollow stick in the forest understory—

without error in the dead of night. Only tiny amounts of light are reaching its light-sensing cells, but it appears to see perfectly well. The fact is that we're still a very long way from understanding how insects see the world.

What's the point of wasps?

I've lost count of the number of times I've been asked this sort of question. Beyond making more copies of itself, there's no "point" in any living thing. Every living thing on planet Earth is connected and the lives they all lead in striving to reproduce drive the processes that make this lump of rock a place of vibrant life. This question and others like it are rooted in the notion that humans are above nature and that every other living creature must be of use to us somehow.

Why are wasps/yellow-jackets attracted to sweet things?

The foraging workers of these social insects subsist on a sweet liquid produced by the developing larvae in the nest. The workers normally get some of this liquid as a reward when they bring prey—other insects—back to the nest. In temperate locations toward the end of the summer when the nest begins to decline, there are fewer developing larvae in the nest and the workers start to seek their sugar hit elsewhere, for example from fallen fruit and sugary drinks.

What's the difference between moths and butterflies?

There's no real difference. Recent studies have shown that butterflies are diurnal moths that diverged from their nocturnal relatives about 100 million years ago, probably to avoid being eaten by bats. The first lepidopteran was a small insect that lived around 300 million years ago in the late Carboniferous period. The adults had mandibles and the larvae probably fed internally on nonvascular land plants.

Are beetles the most diverse animals?

There are certainly more described beetle species than any other animal—currently about one in four animal species is a beetle. Compared with many other groups of insects, beetles have always been more favored by insect collectors too. It's highly likely that the flies (Diptera) and the wasps, ants, and bees (Hymenoptera) may be even more diverse than the beetles. In these groups there's an extraordinary richness of tiny species, most of which are still unknown to science. Beyond the insects, there are also a huge variety of mites and nematodes out there and they're probably among the most diverse animals.

> Social wasps are not out to ruin your day. Cut them some slack.

Do insects sleep?

Diurnal insects "rest" during the night and nocturnal insects "rest" during the day, normally in positions where they are least conspicuous. Solitary bees and wasps can often be found "roosting" as darkness approaches, clinging onto a leaf edge or stem with their mandibles. It's unlikely that insects sleep in the same way as mammals or birds.

Where do insects go in the winter (or when it rains)?

Insects see out the cold winter months in sheltered spots and often in a state of dormancy. These sheltered spots include the upper layers of the soil, beneath bark, logs, and rocks, inside caves, and even in our houses. Depending on the species, it can be the eggs, immature stages, or adults that overwinter, but in all cases, changes in day length and temperature are the triggers.

Why do entomologists have to kill insects?

In most cases, an insect can only be reliably identified by looking at tiny features on its body. This can only be done when the insect is dead and it is not something that entomologists take pleasure in. Museum collections around the world have millions of insect specimens, some of which were collected centuries ago. These collections are important repositories of information. They tell us about our changing world, how and where certain species live, and about our impact on nature to name just a few.

Why can some insects live without their head?

In large animals, such as mammals, the central nervous system consists of a brain and a nerve cord. Insects have a slightly different arrangement. They, too, have a brain and a nerve cord, but they also have mini-brains, clusters of nerve cells called ganglia at intervals along their nerve cord. In some insects this means the body can survive for a while without the head. The prognosis in this situation is not good though. For instance, with no mouthparts it can't eat very easily and before long it will weaken and die.

Why are there so many different types of insect?

Insects are small animals, so there are loads of tiny micro habitats in which they can live. In adapting to these niches, species are generated. They are also generally very fecund and tend to have short or very short generation times, which provides enormous amounts of variation on which natural selection can act. All these things mean that insects speciate very rapidly. Imagine a small plant-feeding insect, the larvae of which feed on leaves. It's not a huge leap for this one species to diversify into many others, for example, species that feed on developing seed pods, species that feed within the leaves, and species that feed within other tissues of the plant. This is happening all the time. In a mere twenty generations, some male crickets lost their ability to sing because their song attracted a parasitoid fly.

> On the facing page I am using a light to attract insects in northern Myanmar. The diversity of insects in places like this is mind-boggling and humbling.

Why are there no insects in the sea?

In essence, insects are terrestrial crustaceans. Using DNA from lots of different arthropods we can build a family tree of these animals that shows insects are nested within the crustaceans. Long ago, around 500 million years ago, a crustacean adapted to a life on land giving rise to all the insects we see around us today. Some insects dwell in the inter-tidal zone and there are even some (for example, *Halobates* sp.) that live out their whole life on the surface of the open ocean, but there are no insects that have returned to the sea to become fully marine. The reasons for this are unknown. Perhaps the ocean is so full of small crustaceans that there are no niches left for hopeful insects to exploit. This might not always be the case though, in 10 million or a 100 million years' time perhaps some of the inter-tidal insects we see today will have given rise to an abundance of marine insects.

Why are there no really big insects?

The way in which insects "breathe," their musculature, and exoskeleton prevent them from becoming big. As an insect gets larger, the system of tubes that conveys gases to and from the tissues becomes more inefficient and the proportion of the body devoted to this system has to increase. Getting oxygen to the tissues, especially the extremities, becomes harder until a physical limit is reached. Hundreds of millions of years ago there were giant arthropods, such as *Arthropleura*—a giant millipede. When these animals were trundling around, the oxygen concentration of the atmosphere was much higher than today, which may have overcome the problem of getting enough oxygen to the tissues. At around the same time, griffinflies flitted through the air. Griffinflies such as *Meganeura monyi* and *Meganeuropsis permiana* had wingspans of around 28 inches (70 cm), but they were less bulky than the chunkiest insects alive today, e.g. some of the beetles and weta. For an insect to grow, the exoskeleton has to be shed; this is a biologically expensive thing to do and very risky. Lots of material is needed to make the new exoskeleton, and after the old one is shed, the animal is soft and very vulnerable. Smaller muscles are also proportionally stronger than bigger muscles, so a really big insect would have trouble even supporting itself. Imagine a 20- or 40-pound (9- or 13.5-kg) insect larva that successfully manages to squeeze out of its old exoskeleton only to be crushed under its own weight before its new exoskeleton fully hardens. Water supports the body of an animal, which is why by far the largest arthropods are aquatic.

What is a species?

This is a tricky question and there's no clear-cut answer. The most widely used definition is based on the biological species concept proposed by Ernst Mayr in 1940: "Species are groups of actually or potentially interbreeding natural populations, which are reproductively isolated from other such groups." This is well-known and has some intuitive appeal, but it's very limited and it is biased toward animals. For example, we cannot apply it to bacteria and many other organisms where asexual reproduction is the norm. With that said, these organisms still exist as discrete groups with shared characteristics, so the definition needs to be tweaked. There are currently 32 different species definitions and there isn't one that applies to all life. We have to remember that the concept of a species is a useful tool, but one that we invented. Our minds thrive on order and we seek this when making sense of the natural world, but nature is dynamic and fuzzy with poorly defined boundaries. Regardless of the organisms in question, lineages are continually splitting and rejoining over time—a bit like the braided flow of a river in a delta—and this is the only commonality we can find.

Are insects heading toward extinction?

No. There have been stark declines in the populations of some insects in some places, but, as a group, they're certainly not heading toward extinction. Indeed, there will be insects long after we've gone. This is no reason to be complacent though. Humans are changing the world at an alarming pace, putting the other species we share this planet with under intense pressure. On the whole, we know so little about insects and their population trends that we just don't know the full extent of our impact on these animals.

What can we do to protect insects?

There are choices we can all make to lessen our impact on the planet and the other species we share it with. First and foremost: buy less stuff, whether it's food, clothes, or gadgets. Everything we buy has an environmental price tag, often a hefty one. The chickens that are reared in their hundreds of millions to supply fast-food outlets are fed soya, which contributes enormously to habitat destruction, especially in South America. The clamor for rare earth metals that are used in our electronic gadgets is driving rampant habitat destruction in far-flung places around the world, many of which are teeming with life. Think about ways in which you can use fewer fossil fuels. Drive and fly less, and walk or cycle more. Where you can, reduce what you use; reuse, and recycle. More locally, try to avoid food that has been produced using copious amounts of pesticides and make your garden and green spaces insect friendly by offering habitats and resources that insects need and thrive in.

> Every passing day sees the loss of more tropical forest—habitats that are bristling with life. First, the roads come. This fragments the forest and makes it much easier for people to access and cut down ever larger areas. These places are the treasures of our planet and we have to wake up to the madness of our actions.

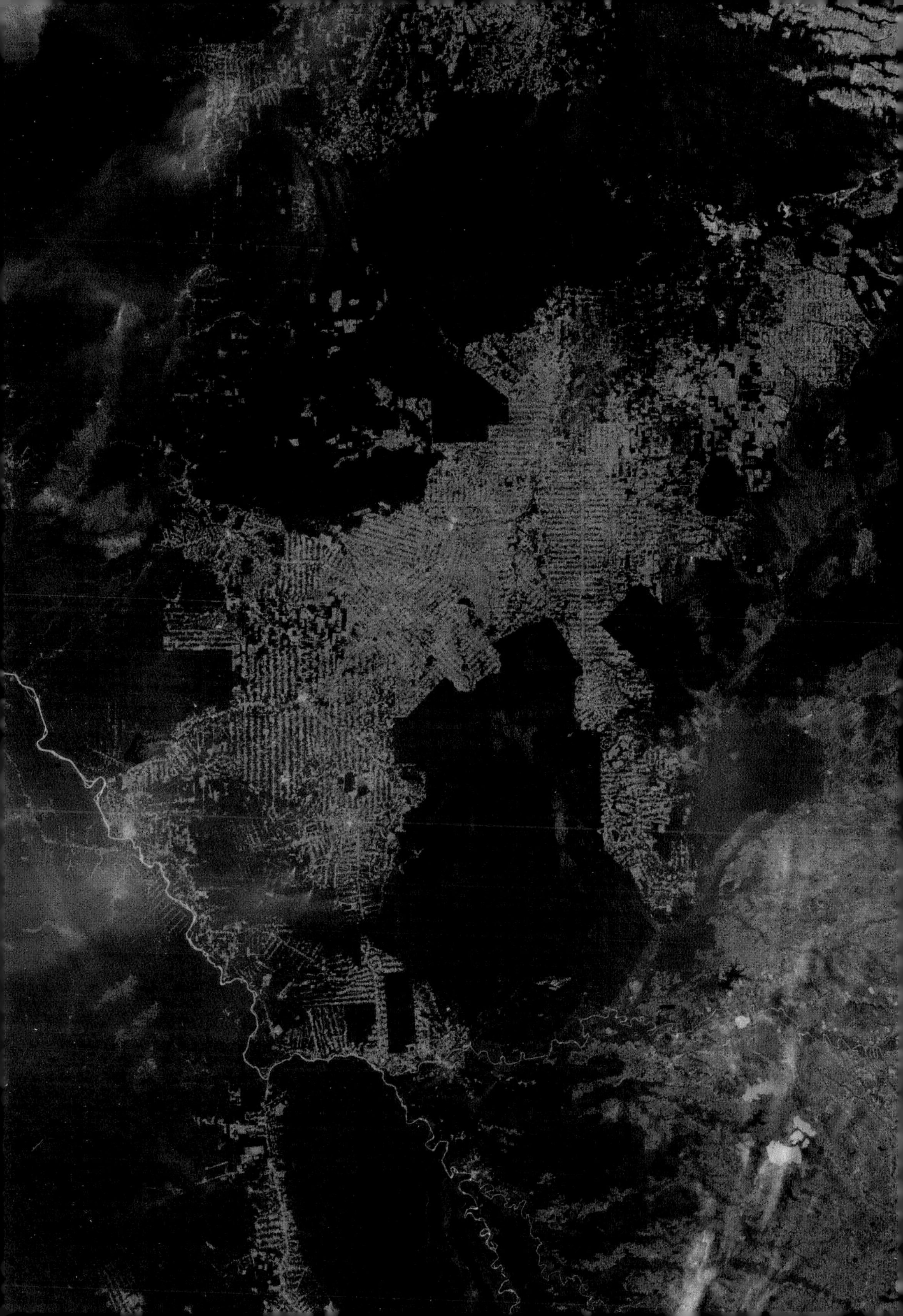

GLOSSARY

Beltian body
Small capsules rich in lipids and proteins found on the tips of leaflets in some acacias.

Book lung
A type of respiratory organ found in certain arachnids.

Bryzoans
Aquatic, sessile animals that are typically colony forming.

Calyx
A funnel-shaped expansion of the vas deferens or oviduct of insects.

Carrion
The decaying flesh of dead animals.

Caste
Individuals within a colony of eusocial animals that are specialized in a particular function, e.g. foraging, defense, or reproduction.

Cephalocarids
Tiny, shrimp-like crustaceans that live in all types of marine sediment.

Coleoptera
The order of insects in which all beetles belong.

Cocoon
A silken case (sometimes reinforced with other material) spun by insect larvae for protection during pupation.

Conopid fly
A fly belonging to the family Conopidae, sometimes commonly known as thick-headed flies.

Crustaceans
A group of arthropods that includes crabs, barnacles, copepods, etc.

Diploid
A cell or an organism that contains two sets of chromosomes, usually one set from the mother and one set from the father.

Diptera
The order of insects in which all true flies belong.

Echolocation
The location of objects by reflected sound.

Elytra
The hardened forewings of certain insects, especially beetles.

Entomologist
A person who studies insects.

Epiphytic plant
A plant that grows on another plant and depends on it for support but not food.

Ethologist
Someone who studies animal behavior.

Eusocial
Any colonial animal species that lives in multigenerational family groups in which most individuals cooperate to aid relatively few (or even a single) reproductive individual.

Exoskeleton
The external supportive covering of all arthropods.

Fossorial
An animal adapted to burrowing.

Frass
Debris or excrement produced by insects.

Gall
An abnormal outgrowth of plant tissue triggered by insects, mites, fungi, viruses, or bacteria.

Ganglia
Plural of ganglion—an encapsulated collection of nerve-cell bodies, usually located outside the brain.

Genome
An organism's complete set of DNA, which include all its genes.

Gongylidia
Hyphal swellings on the fungi cultivated by some insects.

Haplodiploidy
A sex-determination system in which males develop from unfertilized eggs and are haploid, and females develop from fertilized eggs and are diploid.

Haploid
A cell or organism that contains a single set of chromosomes.

Hemocoel
A body cavity of many invertebrates that contains hemolymph.

Hemolymph
The arthropod equivalent of blood in vertebrates.

Honeydew
The sugary liquid produced by sap-feeding insects.

Hymenoptera
The order of insects in which all wasps, ants, and bees belong.

Instar
A developmental stage of arthropods, such as insects, between each molt until sexual maturity is reached.

King ant
Reproductive male ants that mate with new queens.

Labium
A lower mouthpart of an insect.

Lekking
An aggregation of animals gathered to engage in competitive displays and courtship rituals.

Lepidoptera
The order of insects in which moths and butterflies belong.

Media
Medium-sized, generalized foragers in ant colonies, especially leafcutter ants.

Minim
The smallest workers in leafcutter ant colonies which tend to the growing brood or fungus gardens.

Molt
The process in which an insect sheds its exoskeleton; also known as ecdysis.

Nuptial flight
A flight of sexually mature social insects in which mating takes place.

Nuptial gift
Items provided by an insect to a mate prior to mating.

Nymph
An immature form of an insect that does not change greatly as it grows, for example, a shield bug or grasshopper.

Ommatidia
The individual units of an insect's compound eye.

Ootheca
A case or capsule containing eggs.

Ovipositor
A tubular organ through which a female insect deposits her eggs.

Queen
The reproductive female in a colony of eusocial insects.

Parasite
An organism that lives on or in an organism of another species, and from which it obtains food.

Parasitoid
An organism that lives on or in a host organism, and ultimately kills the host.

Pathogen
A biological agent that causes disease or illness to its host.

Pheromone
A chemical substance usually produced by an animal that influences the behavior of another individual of the same species.

Phragmosis
A method of closing the burrow or nest by means of some specially adapted part of the body.

Polyembryonic
The phenomenon where two or more embryos develop from a single fertilized egg.

Polygerm
A cluster of germ cells or morulae found in the polyembryonic development of some insects.

Proboscis
The elongated sucking mouthpart of some insects that is typically tubular and flexible.

Puddle
The aggregation of some insects on wet soil or dung to obtain moisture and nutrients.

Pupa
A stage in the life of an insect between the larva and adult.

Pygidial
The posterior body region or segment of insects.

Raptorial
A limb or other structure adapted for seizing prey.

Remipedes
A rare group of marine crustaceans that have paddle-like appendages on their long trunks.

Setae
Several different bristle-, hair-, or scale-like structures found on insects.

Spermatophore
A package of sperm often with nutritious layers passed from the male to the female during mating.

Spicule
Small, hard, calcareous, or siliceous bodies that serve as the skeletal elements of various aquatic invertebrates.

Symbiotic
Involving interaction between two different organisms living in close physical association.

Tachinid
A fly belonging to the family Tachinidae, all of which are parasitoids or occasional parasites.

Trachea
Fine chitinous tubes in the body of an insect, conveying air to and from the tissues.

Triungulin
The mobile first instar larva of an insect that undergoes hypermetamorphosis.

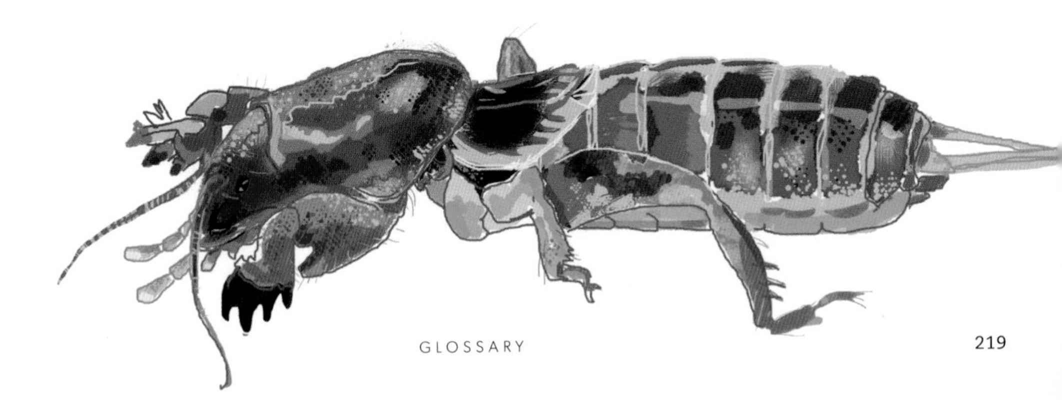

INDEX

Achias rothschildi 41
Adetomyrma 134
Allomerus 204
decemarticulatus 202–3
Amblyopinus 205
ambush predators 73–9
Anabrus simplex 44
ant-decapitating flies 164–5
antennae 8, 13, 23, 166
antlions 73, 78
ants 49, 54, 106, 117, 143
  army ants 136
  and blue butterflies 154–5
  bullet ants 144, 204
  Dracula ants 134
  driver ants 136
  eggs 24
  eusociality 123, 124, 125, 128, 130–41, 144, 148–50
  fire ants 144
  flying ants 139
  and fungi 132, 136, 186, 202–3
  leafcutter ants 24, 53, 93, 125, 130, 132, 136–7, 186
  lemon ants 195
  Maricopa harvester ants 144
  plant mutualisms 195–204
  pollination 69
  raspberry crazy ants 144
  red ants 154–5
  sap sucking insects 201
  slave-making ants 141
  trap-jaw ants 144
  wood ants 134, 135, 140, 144
aphids 24–5, 32, 37, 63, 64, 123, 124, 126–7, 134, 201, 212
Aphomia sociella 99
Apis mellifera 146
aposematic coloration 62, 116, 117
appendages 6, 8, 13, 15
aquatic insects 61, 86, 215
Arachnocampa luminosa 74
Argentine ants 140
army ants 136
Ascalaphidae 73
assassin bugs 75, 105, 117, 194
asynchronous muscle 17
Attelabidae 50
Austroplatypus incompertus 185

bacteria 192–3, 199
bark beetles 185
Batesian mimicry 107
bats, and moths 97
beaded lacewings 87
bed bugs 46–7
bee flies 32, 36
bee-grabber flies 49
bees 36, 54, 117, 162, 178
  cuckoo bees 103
  eusociality 123, 124, 128, 145, 148
  pollination 68, 69
  solitary 36, 45, 214
  sweat bees 212–13
  see also honey bees
beetles 6, 10, 13, 21, 63, 89, 149–50
  ambush predators 74
  bark beetles 185
  bombardier beetles 118, 150
  burying beetles 54, 89
  carrion beetles 117
  click beetles 74
  colors 102
  courtship 39
  decaying wood 91
  diversity 213
  dung beetles 52–3, 70, 87, 90, 92–3
  eggs 49
  elytra 18
  Epomis 76–7
  eusociality 123, 124
  fungi farming 185–6
  fungus beetles 54
  ironclad beetles 111
  jewel beetles 102
  leaf beetles 63, 104
  leaf miners 65
  lily beetles 12, 104
  longhorn beetles 112, 117, 119
  mandibles 112
  oil beetles 24, 36
  parasitoids 161
  parental care 50–4
  pollination 69
  reed beetles 61
  rove beetles 39–40, 75, 78, 109, 117, 149, 152, 205
  ship-timber beetles 186
  snail predation 87
  telephone-pole beetles 32, 37
  tiger beetles 74, 82
  tortoise beetles 50–1, 104
  water beetles 114
beewolf 54–7, 103, 192
bella moths 44
Beltian bodies 134, 199
biomass decline 206–7
Bledius 51–2
blow flies 89
blue butterflies 148, 153–5
body size 215
bombardier beetles 118, 150
bone-house wasps 105
bone-skipper flies 89
bot flies 49, 176, 177
brain 18, 80, 132, 165, 168, 214
breathing 215
bullet ants 144, 204
bull horn acacias 198–9
burying beetles 54, 89
butterflies 13, 15, 19, 100, 148
  blue butterflies 148, 153–5
  cabbage white butterflies 43, 47
  clearwing swallowtails 47
  decline 206–7
  mating 47
  moth difference 213
  pollination 69
  red postman butterflies 47
  scales 114

cabbage white butterflies 43, 47
caddisflies 27, 112
Calleremites subornata 100
Calyptra 180
calyx 191
camouflage 73, 76, 100–2
carnivores 72–87
carrion beetles 117
Cassidinae 50–1
castes 135–8
caterpillars 6, 19, 63, 79, 89, 126, 170–1
  blue butterflies 153–5
  fecal mimicry 104
  mimicry 106
  parasitoids 161
  spines 112
chemicals 105, 149–50, 155
  communication 125
  mimicry 103
  plant defenses 62–3, 65, 115–16, 166
  repellent smells 116–17
  slave-making ants 141
  stenusin 109
  toxic semen 44, 47
  toxic vapors 87
  warning colors 102
  weapons 115–19, 142, 143
  see also venom

chemoreceptors 22–3
Chrysopidae 49
cicadas 22, 38, 161
Cimex 46–7
claws 112
clearwing swallowtails 47
click beetles 74
climate change 209
clones 25, 37, 126, 172–3
cockroaches 49, 168–9
color 12–13
  aposematic 62, 116, 117
  disguise 102
compound eyes 8, 23–4
Conopidae 49, 162
conopid flies 49, 162
convergent evolution 66, 73, 78, 83, 108, 124
Coprophanaeus lancifer 53
Cordia trees 204
courtship 38–40, 42–3, 75, 99
coxa 13
Cressida cressida 47
cricket flies 99
crickets 15, 22, 39, 44, 99, 106, 202–3
cuckoo bees 103
cuckoo wasps 103

damselflies 45, 78
Dasymutilla gloriosa 108
deer flies 175
Dermatobia hominis 49
Deuteragenia ossarium 105
dormancy 214
Dracula ants 134
dragonflies 45, 78
driver ants 136
Drosophila melanogaster 47
dung beetles 52–3, 70, 87, 90, 92–3

ears 22
earwigs 50
echolocation 49, 97, 166
eggs 24–5, 32–3, 35–6, 46, 49
  eusocial insects 135–6
  parasitoids 160, 164, 166–7, 169–71, 172, 191
  parental care 50–7
elytra 18
emerald cockroach wasps 168
Epomis beetles 76–7
Eucoeliodes mirabilis 104–5
eumenid wasps 171
Eupithecia 79
European honey bees 146
eusociality 120–55, 185
  castes 135–8
  colony defense 142–6
evolution 40, 41, 99
  convergent 66, 73, 78, 83, 108, 124
  counteradaptations 47, 63, 119
  wings 15
exoskeleton 6, 8, 10, 12–13, 34, 111–14, 215
extinction threat 206–9, 216
eyes 8, 23–4, 212
eye stalks 41

fairy wasps 18–19
feces 49, 90, 162, 200
  defense 51, 104, 143
  mimicry 104–5
  see also dung beetles
femur 15
fig wasps 35
fire ants 144
fireflies 74–5
flea beetles 15, 109
fleas 109, 175–6, 205
flies 17, 63, 89, 92–3, 152
  ant-decapitating flies 164–5
  bee flies 32, 36
  blow flies 89
  bone-skipper flies 89
  bot flies 49, 176, 177
  colors 102
  conopid flies 49, 162
  cricket flies 99
  decaying wood 91
  deer flies 175
  fruitflies 47
  horse flies 175
  hoverflies 15, 17
  leaf miners 65
  life spans 212
  owlflies 73
  parasites 174–7
  pollination 69
  small-headed flies 162
  tachinid 166, 171
  timber flies 91
  tsetse flies 50
  warble flies 176–7
flight 15–18, 109
  flying ants 139
  tandem flying 45
flying ants 139
Formica 135
  yessensis 140
formic acid 144, 195
fruitflies 47
fungi 91, 127
  ant farming 132, 136, 186
  beetle farming 185–6
  trap ants 202–3
fungus beetles 54
fungus gnats 74
fungus weevils 41

galls 63, 66–7, 126–7
giant honey bees 145
Girault, A. A. 159
glowworms 39
grasshoppers 22, 38–9, 63, 100
griffinflies 215
growth 10, 12

habitat loss 208–9
halteres 17
hearing 22
Hedychrum rutilans 103
Heliconius erato 47
hemolymph 15, 34, 134, 170, 188, 190
herbivores 61–7
Hirtella 204–5
  physophora 203
Homoeocera albizonata 114
honey bees 18, 55–7, 135, 138, 145–6
  eusociality 123
  waggle dance 28
honeydew 64, 134, 175, 201, 204
hornets 146, 152
horse flies 175
horsehair worms 177
host manipulation 162–3, 168–9
hoverflies 15, 17
hunting 80–7
hypermetamorphosis 19
hyperparasitoids 159

ichneumon wasps 159, 171
imaginal discs 21
inchworms 79
infrared detection 21
insecticide resistance 25, 27
ironclad beetles 111

Japanese honey bees 146
Japanese hornets 146
jewel beetles 102
jumping 109

katydids 22, 99, 100

lacewings 49, 87, 105, 106
ladybirds 116, 163
larvae 63, 105, 138, 162, 169, 170, 212
  ambush predators 73–4, 77–8
  aquatic hunters 86
  beewolf 193
  caddisfly 112
  decaying wood 91
  leaf miners 65
  life spans 212
  metamorphosis 19

parasitoids 164–5, 172, 192
parental care 50–7
ship-timber beetles 186
solitary wasps 188–9
Strepsiptera 178–9
telephone-pole beetles 37
see also caterpillars maggots

leaf beetles 63, 104
leafcutter ants 24, 53, 93, 125, 130, 132, 136–7, 186
leaf hoppers 15, 109, 201
leaf-miners 63, 65
leaf-rolling weevils 50
learning 27–9
Leistotrophus versicolor 39–40
lekking 42
lemon ants 195
Leptanilla japonica 134
lesser water boatman 39
lichen katydids 100
life spans 212
lily beetles 12, 104
Linepithema humile 140
Lithinus rufopenicillatus 100
Lobocraspis griseifusa 180
Lomamyia latipennis 87
longhorn beetles 112, 117, 119
long-tailed moon moths 97
Lymexylidae 186

maggots 89, 177
mandibles 112
mantises 49
orchid mantises 75–6
praying mantises 78
mantisflies 78
Maricopa harvester ants 144
Markia hystrix 100
mate guarding 45
mate selection 39–40, 42
mating 45–7
matricide 37
Mayr, Ernst 216
Megalopta genalis 212–13
Meganeura monyi 215
Meganeuropsis permiana 215
metamorphosis 19–21
microbes 64
Micromalthus debilis 37
midges 17, 66
migrations 15
millipedes 91
mimicry 100–8, 149
acoustic 103
chemical 103
fecal 104–5
other animals 106–7
miniaturization 18–19
mites 188–90, 205
mole crickets 15
molt 10, 12, 37
mormon crickets 44
mosquitoes 49, 174, 175
moss-mimic stick insects 100
moths 15, 19, 22, 89
acoustic mimicry 103
and bats 97
bella moths 44
butterfly difference 213
camouflage 100
fecal mimicry 104
leaf miners 65
long-tailed moon moths 97
parasites 175, 180–1
parasitoids 161
pollination 69
scales 97, 114
silk 112–13
tiger moths 97, 114
wax moths 99
Müllerian mimicry 107
multi-wasps 172–3
Myrmecodia 197
Myrmeleontidae 73
Myrmica 154–5

navigation 54, 81, 93, 212
Nemopteridae 78
nervous system 18, 21, 214
nuptial gifts 43–4
Nylanderia fulva 144
nymphs 33–4, 78, 97, 105, 161, 212

oil beetles 24, 36
ommatidia 23
ootheca 49
orchid mantises 75–6
Ormia ochracea 22, 99
owlflies 73

Paederus 117
pain 212
palps 13
Pameridea roridula 194
parasites 148–50, 154–5, 174–81
parasitoid flies 22, 136, 162, 164–6
parasitoids 156–81
parasitoid wasps 18–19, 49, 62, 159–60, 166–73
host manipulation 163
and viruses 191–2
parental care 50–7
parthenogenesis 25
pederine 117
Perisceptis carnivora 79
pesticides 209
Phasmatodea 49
Phengaris 153
arion 155
pheromones 15, 38, 42, 45, 47, 146
Philanthus 55–6
Phoreticovelia disparata 44
photoreceptors 23
phragmosis 143
Pieris rapae 47
pitcher plants 200
planthoppers 109, 161
plants
defenses 62, 65, 115–16, 166
insect herbivores 61–7
insect mutualisms 194–205
see also pollination
pollination 35, 46, 68–71, 93, 184
polyembryonic development 172–3
polygerm 172
praying mantises 78
predators 72–87
Psychopsis mimica 106
pupa 21, 34
pupation 112–13

Quedius dilatatus 152

raspberry crazy ants 144
red ants 154–5
red postman butterflies 47
reed beetles 61
Regimbartia attenuata 114
reproduction 24–7, 32–3, 35–6
bark beetles 185
eusocial insects 135–6
life cycles 30–57
mate selection 39–40, 42
mating 45–7
parasitoids 162–71
ship-timber beetles 186
solitary wasps 188–90
resilin 109
resin assassin bugs 75
ritual combat 41
Roridula 194
rove beetles 39–40, 75, 78, 109, 117, 149, 152, 205

sand wasps 108
sawflies 63, 65, 170–1
scale insects 201, 204
scales 13, 97, 114, 155
scavengers 88–93
screwworms 177
seaweed 91
seed dispersal 70, 92
senses 21–4
sensory cells 21
setae 21
shield bugs 51
ship-timber beetles 186
shore earwigs 117
silk 112–13
simple eyes 8, 23
slave-making ants 141
sleep 214
small-headed flies 162
smell 22–3, 38
blue butterflies 154, 155
repellent 116–17
snail predators 87
'sneaky males' 39–40
Solenopsis 144
solitary bees 36, 45, 214

solitary wasps 54, 69, 124, 128, 214
and microbes 192–3
and mites 188–90
song 38–9, 99
species, definition 216
spermatophores 43–4
spines 161
spittlebugs 97
spongeflies 86
Stenus 78, 109
stenusin 109
Sthenauge parasiticus 161
stick insects 49, 100
stink bugs 116–17
stone grasshoppers 100
Strepsiptera 177, 178–9
stridulation 39, 44
supercolonies 140–1
superorganisms 125
swarms 42
sweat bees 212–13
symbiosis
insects and fungi 132, 136, 185–6, 202–3
microbes 64, 117, 175, 191–3

tachinid flies 166, 171
tandem flying 45
tannins 62
tarsus 15
taste 22, 38
Teleogryllus oceanicus 99
telephone-pole beetles 32, 37
Temnothorax 132
teneral state 34
termites 74, 87
eggs 24
eusociality 123, 124, 125, 130, 135–6, 138, 142–3, 148–9
microbiota 193
thermoregulation 108
thrips 25, 123, 124, 126–7
thynnine wasps 45–6
Thyreophora cynophile 89
tibia 15
tiger beetles 74, 82
tiger moths 97, 114
timber flies 91
tortoise beetles 50–1, 104
Trachypetrella 100
trap-jaw ants 144
traumatic insemination 46–7
treehoppers 100, 104, 201
trigonalid wasps 170–1
triungulins 36–7
trochanter 13
true bugs 21, 22, 106
Trychopeplus laciniatus 100
tsetse flies 50

Utetheisa ornatrix 44

velvet ants 108
venom 117, 144, 167, 168–9
Vespa 152
crabro 152
mandarinia 146
viruses 167, 191–2
Volucella inanis 152
von Frisch, Karl 28

waggle dance 28
Waitomo Caves 74
warble flies 176–7
warning colors see aposematic coloration
wasps 27, 54, 117, 178–9
bone-house wasps 105
clones 172–3
cognition 28
cuckoo wasps 103
emerald cockroach wasps 168
eumenid wasps 171
eusociality 123, 124, 128, 145, 148, 152
fairy wasps 18–19
fig wasps 35
galls 66
ichneumon wasps 159, 171
mimicry of 107
point of 213
pollination 69
sand wasps 108
sweet attraction 213
thynnine wasps 45–6
trigonalid wasps 170–1
see also parasitoid wasps; solitary wasps
water beetles 114
water bugs 78
wax moths 99
web spinners 113
weevils 100, 104–5
armor 111
fungus 41
leaf-rolling 50
wings 8, 13, 15–18, 42, 109, 176
wood ants 134, 135, 140, 144
wood decay 90–1
worm-lions 73

Zeus bugs 44
Zopherus nodulosus 111

ACKNOWLEDGEMENTS

First and foremost, a huge thank you to my editor and picture researcher, Natalia Price-Cabrera, for pulling all of this together and resisting my demands to include every single photo of insects I could get my hands on. Thank you to commissioning editor, Kate Shanahan, and publisher, Nigel Browning at UniPress Books for commissioning me in the first place. Thanks also to my family for giving me the time to work on this alongside my other commitments in what has been the strangest of years and continues to be. Thanks to the entomological community on Twitter for providing input in one way or another, Zenobia Lewis, illustrator Sarah Skeate, book designer Paul Palmer-Edwards, and all of the contributors who have provided images and been so generous with their time and knowledge. Each one of the images in this book has a story and when you look at them, spare a thought for the patience and skill that goes into finding the subject and photographing it. Let me also give a big shout-out to all the scientists and naturalists, past and present, who have dedicated years, decades, and whole careers to understanding insects and their complex lives.

You can find more about my work and the incredible lives of insects at: https://www.rosspiper.net/

PICTURE CREDITS

Every effort has been made to trace copyright holders and acknowledge the images. The publishers welcome further information regarding any unintentional omissions.

All images are courtesy of the author, Dr Ross Piper, unless otherwise indicated below:

p2: Courtesy of tcareob72/Shutterstock. p4–5: Courtesy of Uditha Wickramanayaka/Shutterstock. p7: Reproduced and modified with permission from Gonzalo Giribet and Greg Edgecombe. p8–9: Courtesy of Protasov AN/Shutterstock. p10 (top), p11: Courtesy of Takuya Morihisa. p10 (bottom): Courtesy of Melvyn Yeo. p12 (bottom): Courtesy of Udo Schmidt (CC BY-SA 2.0). p16: Courtesy of Hans Pohl. p20: Courtesy of Protasov AN/Shutterstock. p30–31: Courtesy of Paul Bertner. p32: Courtesy of Melvyn Yeo. p33: Courtesy of Eric Isselee/Shutterstock. p34: Courtesy of Böhringer Friedrich, edited by Fir0002 (CC BY-SA 2.5). p36: Courtesy of ozgur kerem bulur/Shutterstock. p38 (left): Courtesy of Sarefo (CC BY-SA 2.0); (right): Courtesy of ExaVolt (CC BY-SA 4.0). p39: Courtesy of WUT. ANUNAI/Shutterstock. p41 (top): Courtesy of Melvyn Yeo; (bottom): Courtesy of Worraket/Shutterstock. p42: Courtesy of Mladen Mitrinovic/Shutterstock. p43 (top): Courtesy of Paul Bertner; (bottom): Courtesy of Vinicius R. Souza/Shutterstock. p44 (bottom left): Courtesy of Dan Olsen/Shutterstock; (centre right): Courtesy of Dan Johnson. p45: Courtesy of YOSHIZAWA Kazunori. p46: Courtesy of Melvyn Yeo. p47: Courtesy of Greg Hume (CC BY-SA 3.0). p48 (bottom): Courtesy of Victor Engel. p52: Courtesy of Melvyn Yeo. p53 (bottom): Courtesy of Nick Royle. p54 (right): Courtesy of Will Hawkes. p56: Courtesy of thatmacroguy/ Shutterstock. p57: Courtesy of Prof. Dr. Martin Kaltenpoth. p60, 62: Courtesy of Cathy Keifer/Shutterstock. p63: Courtesy of Frank Canon. p64: Courtesy of Melvyn Yeo. p65: Courtesy of Protasov AN/ Shutterstock. p66 (left): Courtesy of D. Kucharski K. Kucharska/ Shutterstock; (right) Courtesy of colin robert varndell/Shutterstock. p67 (bottom): Courtesy of D. Kucharski K. Kucharska/Shutterstock. p69: Photo by Hannier Pulido, courtesy of the De Moraes and Mescher Laboratories. p70: Courtesy of Michael & Patricia Fogden/ Minden Pictures/Alamy Stock Photo. p71: With thanks to Prof. Jeremy J. Midgley and Dr Joseph DM White/Photo © Dr Joseph DM White/Midgley et al. 2015. p72: Courtesy of Marcio Cabral/ BIOSPHOTO/Alamy Stock Photo. p74: Courtesy of Peter Yeeles/ Shutterstock. p75: Courtesy of Melvyn Yeo. p76: Courtesy of Kawin Jiaranaisakul/Shutterstock. p77: Courtesy of Alex Shlagman. p81: Courtesy of Melvyn Yeo. p82: Courtesy of Lyle Buss. p85: Courtesy of Matt Bertone. p86 (left): Courtesy of Jam Hamrsky; (right): Courtesy of Henri Koskinen/Shutterstock. P87: Courtesy of Paul Bertner. p88: Courtesy of 7th Son Studio/Shutterstock. p90: Courtesy of Paul Bertner. p92: Courtesy of Rudmer Zwerver/Shutterstock. p93: Courtesy of Márcio Araújo. P94–95, 96: Courtesy of Paul Bertner. p97: Courtesy of Thomas R. Neil. p98: Courtesy of Matee Nuserm/ Shutterstock. p99 (top): Courtesy of Jpaur (CC BY-SA 3.0). p100: Courtesy of Ragnhild&Neil Crawford (CC BY-SA 2.0). p101 (top): Courtesy of Paul Bertner; (bottom) Courtesy of Melvyn Yeo. p102: Courtesy of Andy Murray. p103 (top right): Courtesy of Melvyn Yeo. p104: Courtesy of Larry Minden/Minden Pictures/Alamy Stock Photo. p105 (top): Courtesy of Tyler Fox/Shutterstock. p106: Courtesy of Jay Ondreicka/Shutterstock. p108 (top): Courtesy of Joseph Wilson; (bottom) Courtesy of Melvyn Yeo. p109: Courtesy of Udo Schmidt (CC BY-SA 2.0). p112, 113 (top): Courtesy of Andy Murray. p113 (bottom): Courtesy of Melvyn Yeo. P119 (top): Courtesy of Bernard Dupont (CC BY-SA 2.0); (bottom) Courtesy of Arthur V. Evans. p120–121: Courtesy of Stastny_Pavel/Shutterstock. p122: Courtesy of Paul Bertner. p123: Courtesy of RAMLAN BIN ABDUL JALIL/Shutterstock. p125 (left): Courtesy of Pavel Krasensky/ Shutterstock; (right) Courtesy of Noah Elhardt (CC BY-SA 4.0).

p126: Courtesy of Mayako Kutsukake. p127: Courtesy of Dr. Harunobu Shibao. P128–129 (bottom): (c) Heather Holm/www. pollinatorsnativeplants.com. p131: Courtesy of Michael Siluk/ Shutterstock. p133: Photo taken by M. Bollazzi; from Bollazzi et al., Insectes Sociax (2012) 59:487–498/Courtesy of Prof. Dr. Flavio Roces. p134: Courtesy of Piotr Naskrecki/Minden Pictures Contributor: Minden Pictures/Alamy Stock Photo. p135: Courtesy of Matt Bertone. p136: Courtesy of Dr Morley Read/Shutterstock. p137 (top): Courtesy of Paul Bertner; (bottom) Courtesy of nikjuzaili/ Shutterstock. p138: Courtesy of Mirko Graul/Shutterstock. p139 (left): Courtesy of Jeremy Christensen/Shutterstock; (right) Courtesy of Dragomir Radovanovic/Shutterstock. p141 (left): Courtesy of Adrian A. Smith (CC BY 2.5); (right) Courtesy of James C. Trager (CC BY-SA 3.0). p142: Courtesy of Paul Bertner. p144 (left): Courtesy of Minsoo Dong (CC BY-SA); (right) Courtesy of Smartse (CC BY-SA 3.0). p145: Courtesy of Nireekshit (CC BY-SA 4.0). p147 (top): Courtesy of Takahashi (CC BY-SA 3.0); (bottom) Courtesy of Yasunori Koide (CC BY-SA 4.0). p148: Courtesy of Bruce Blake (CC BY-SA 4.0). p149 (top left): Courtesy of Taisuke KANAO; (bottom left): Courtesy of Andy Murray (CC BY-SA 2.0); (bottom right) Courtesy of KAKIZOE Showtaro. p150, p151 (top): Courtesy of Taisuke KANAO. p151 (bottom): Courtesy of Dr. Takashi Komatsu. p152 (top): Courtesy of Gail Hampshire (CC BY 2.0); (bottom) Courtesy of Udo Schmidt (CC BY-SA 2.0). p153: Courtesy of Charles J Sharp (CC BY-SA 4.0). p156–157: Courtesy of Young Swee Ming/ Shutterstock. p158: Courtesy of Beatriz Moisset (CC BY-SA 4.0). p159: Courtesy of Vitalii Hulai/Shutterstock. p160 (top), p161: Courtesy of Melvyn Yeo. p164: Courtesy of Dr. Takashi Komatsu. p165: Courtesy of S.D. Porter, USDA-ARS. p166 (left): Courtesy of Tristram Brelstaff (CC BY 3.0); (right) Courtesy of Beatriz Moisset (CC BY-SA 4.0). p167 (top): Courtesy of Takahashi (Wiki Commons/ Public Domain); (bottom): Courtesy of Wellcome Images/Andrew Polaszek/Natural History Museum/Attribution 4.0 International (CC BY 4.0). p169: Courtesy of P.F.Mayer/Shutterstock. p171: Courtesy of Kristi Ellingsen. p175: Courtesy of Jim Gathany/Centers for Disease Control and Prevention's Public Health Image Library. p177: Courtesy of KAISARMUDA/Shutterstock. p178, p179 (bottom right): Courtesy of Hans Pohl. p181 (top): Courtesy of Leandro Moreas; (bottom): Courtesy of Dumi (CC BY-SA 3.0). p182–183: Courtesy of Somyot Mali-ngam/Shutterstock. p184: Courtesy of Dejen Mengis/USGS Bee Inventory and Monitoring Lab from Beltsville, Maryland, USA. p185: Courtesy of Nikolas_profoto/ Shutterstock. p186 (top): Courtesy of wjarek/Shutterstock; (bottom): Courtesy of Ivan Azimov 007/Shutterstock. p189: Courtesy of Wen-Chi Yeh. p190: Courtesy of Syuan-Jyun Sun (CC BY-SA 4.0). p192 (top): Courtesy of Prof. Dr. Martin Kaltenpoth and Johannes Kroiss; (bottom): Courtesy of Erhard Strohm. P194: Courtesy of Jan Thomas Johansson. P195: Courtesy of Erutuon (CC BY-SA 2.0). p196: Courtesy of Morley Read/Alamy Stock Photo. p197 (top): Courtesy of Vojtěch Zavadil (CC BY-SA 3.0). p198: Courtesy of Dan Perlman/EcoLibrary.org. p199: Courtesy of kafka4prez (CC BY-SA 2.0). p200: Courtesy of D. Magdalena Sorger. p201: Courtesy of Melvyn Yeo. p203: Courtesy of Alain Dejean and Jerome Orivel/ Photo © Alain Dejean. p204: Courtesy of Paul Bertner. p209: Courtesy of John McColgan, Bureau of Land Management, Alaska Fire Service/Photo by John McColgan/USDA. p211: Courtesy of NASA. p213: Courtesy of Bugboy52.40 (CC BY-SA 3.0). p217: Courtesy of Jesse Allen and Robert Simmon/NASA Earth Observatory.